Conference Proceedings of the Society for Experimental Mechanics Series

Series Editor

Kristin B. Zimmerman, Ph.D.
Society for Experimental Mechanics, Inc.,
Bethel, CT, USA

The Conference Proceedings of the Society for Experimental Mechanics Series presents early findings and case studies from a wide range of fundamental and applied work across the broad range of fields that comprise Experimental Mechanics. Series volumes follow the principle tracks or focus topics featured in each of the Society's two annual conferences: IMAC, A Conference and Exposition on Structural Dynamics, and the Society's Annual Conference & Exposition and will address critical areas of interest to researchers and design engineers working in all areas of Structural Dynamics, Solid Mechanics and Materials Research.

More information about this series at http://www.springer.com/series/8922

Shuman Xia • Allison Beese • Ryan B. Berke
Editors

Fracture, Fatigue, Failure and Damage Evolution, Volume 3

Proceedings of the 2020 Annual Conference on Experimental and Applied Mechanics

Springer

Editors
Shuman Xia
Engg Georgia Tech, MRDC 4103
Georgia Inst of Tech, Mechanical
Atlanta, GA, USA

Allison Beese
Materials Engineering Department
Pennsylvania State University
Penn St Univ, PA, USA

Ryan B. Berke
Mechanical and Aerospace Eng
Utah State University
Logan, UT, USA

ISSN 2191-5644 ISSN 2191-5652 (electronic)
Conference Proceedings of the Society for Experimental Mechanics Series
ISBN 978-3-030-60958-0 ISBN 978-3-030-60959-7 (eBook)
https://doi.org/10.1007/978-3-030-60959-7

This Springer imprint is published by the registered company Springer Nature Switzerland AG
The registered company address is: Gewerbestrasse 11, 6330 Cham, Switzerland

Preface

Fracture, Fatigue, Failure and Damage Evolution represents one of seven volumes of technical papers to be presented at the SEM 2020 SEM Annual Conference & Exposition on Experimental and Applied Mechanics organized by the Society for Experimental Mechanics scheduled to be held in Orlando, FL, September 14–17, 2020. The complete proceedings also include volumes on: *Dynamic Behavior of Materials; Mechanics of Time Dependent Materials; Advancement of Optical Methods & Digital Image Correlation in Experimental Mechanics; Mechanics of Biological Systems and Materials, Micro- and Nanomechanics & Research Applications; Mechanics of Composite, Hybrid & Multifunctional Materials; and Thermomechanics & Infrared Imaging, Inverse Problem Methodologies and Mechanics of Additive & Advanced Manufactured Materials.*

Each collection presents early findings from experimental and computational investigations on an important area within experimental mechanics, fracture and fatigue being one of these areas.

Fatigue and fracture are two of the most critical considerations in engineering design. Understanding and characterizing fatigue and fracture has remained as one of the primary focus areas of experimental mechanics for several decades. Advances in experimental techniques, such as digital image correlation, acoustic emissions, and electron microscopy, have allowed for deeper study of phenomena related to fatigue and fracture. This volume contains the results of investigations of several aspects of fatigue and fracture such as microstructural effects, the behavior of interfaces, the behavior of different and/or complex materials such as composites, and environmental and loading effects. The collection of experimental mechanics research included here represents another step toward solving the long-term challenges associated with fatigue and fracture.

Atlanta, GA, USA — Shuman Xia
State College, PA, USA — Allison Beese
Logan, UT, USA — Ryan B. Berke

Contents

Chapter 1
Influence of Fracture and Delayed Effects on Steel-Concrete Composite Structures

Fery Léo, Gwozdziewicz Piotr, and Moutou Pitti Rostand

Abstract This work investigates the fracture behavior of preflex beams submitted to time-dependent effects and long-term loadings. However, it is necessary to consider delayed effects such as shrinkage and creep on these structures because their influence is not negligible. The different materials used in mixed construction are presented. The flowchart applied to compute creep and relaxation effects versus the time is also described. It is interesting to note that the analytical results obtained for a preflex beam are very similar to those obtained using finite element modeling on the LUSAS software. It is also observable that the cracking of the concrete that encases the lower fiber, due to creep, causes an increase of stresses in the steel profile, which can break the beam. The use of reinforcement on the preflex beam decreases the crack opening after 100 years loadings.

Keywords Steel-concrete composite structure · Creep · Shrinkage · Fracture · Preflex beam

1.1 Introduction

Mixed construction is a method which knows a quick development and takes advantage of two materials: steel and concrete [1, 2]. They are linked together by connectors, so that concrete works in compression and steel in tension. This results in a composite beam with improved mechanical characteristics because both concretes are supposed to work in compression. However, it is necessary to consider the impact of delayed effects such as shrinkage and creep on the fracture processes of these structures because their influence is not negligible.

The literature background shows that some authors discussed about the behavior and the manufacturing process of preflex beams [3, 4]. Foort [5] studied the behavior of connectors in preflex beams due to normal and rupture force. Jerome [6] investigated the cracking of stressed preflex beams by comparing numerical and analytical results given by the Ghali and Fabre method [7]. But the impact of delayed effects as creep and relaxation of crack process have not been investigated in this chapter.

The different materials used in mixed construction are presented. At the same time, the conception step of the preflex beam and the numerical flowchart to obtain crack opening are described. It is also observable that the cracking of the concrete that encases the lower fiber, due to creep, causes an increase of stresses in the steel profile, which can lead to the total collapse of the beam.

F. Léo
Université Clermont Auvergne, CNRS, Institut Pascal, Clermont-Ferrand, France

G. Piotr
Facukty of Civil Engineering, Politechnika Krakowska, Warszawska, Krakow, Poland
e-mail: pgwozdziewicz@pk.edu.pl

M. P. Rostand (✉)
Université Clermont Auvergne, CNRS, Institut Pascal, Clermont-Ferrand, France

Institut de Recherche Technologique, CENAREST, Libreville, Gabon
e-mail: rostand.moutou_pitti@uca.fr

S. Xia et al. (eds.), *Fracture, Fatigue, Failure and Damage Evolution, Volume 3*, Conference Proceedings of the Society for Experimental Mechanics Series, https://doi.org/10.1007/978-3-030-60959-7_1

1.2 Material and Methods

1.2.1 Materials: Preflex Beam

The preflex beam is studied in the rest of this chapter. An isostatic beam with a span of 17 m was considered subject to a variable operating load, Fig. 1.1a. The section consists of an HEM 700, first phase concrete B1 (in red) and second phase concrete B2 (in blue), as can be seen in Fig. 1.1b. All dimensions are given here in meters.

According to Table 1.1a, a shade of 275 MPa has been chosen. The elastic moment of the profile is 2299 kN.m. Concrete B1 and B2 are C50/60 and C25/30, respectively, as posted in Table 1.1b.

For numerical computation, a load of 7 kN/m^2 applied to the entire upper surface of concrete B2 is applied. The relative humidity of the environment has been assumed to be 70%.

1.2.2 Modelling Steps

The different steps of the building are posted in Table 1.2. Concrete B1 is poured on day 1 and concrete B2 on day 31. In fact, we wait 28 days after pouring B1 before releasing the prebending forces and 28 days after pouring B2 before commission the beam.

The prebending force is composed of two point forces: one located at the quarter of the beam and another located at three-quarters. It is necessary that the profile does not exceed its elastic limit during the prebending step.

In order to calculate the rupture load of the preflex beam in both cases (with and without the delayed effects), we will increase the operating load until the rupture stress of the profile is reached. The finite element software LUSAS is used in order to implement the numerical model. An important assumptionwhich consists in supposing that concrete B1 does not work anymore and can be removed when it is completely cracked. It should be noted that concrete B1 cracks because the loading is important. In general, this concrete is used in preflex beams for its toughness in compression. When the complete cracking appears, we denote two load cases: one before the total cracking and one after. For the second load case (after cracking), we remove concrete B1 and add a linear force representing the self-weight of B1 before restarting the calculation. Thus, for each load case (the load is gradually increased), we proceeded as detailed in the flowchart in Fig. 1.2, which is as follow:

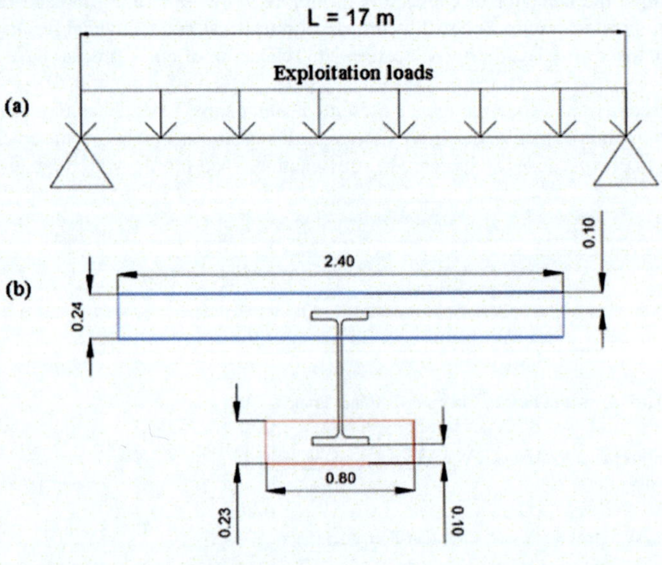

Fig. 1.1 (a) Static model of the beam. (b) Section of beam studied

Table 1.1 (a) Characteristic of the profile HEM 700. (b) Characteristics of the concrete in B1/B2 phases

(a)

Mechanical characteristic of steel		
HEM 700		
A =	383	cm^2
h =	716	mm
b =	304	mm
t_f =	40	mm
t_w =	21	mm
y cdg =	358	mm
A_{vz} =	169,8	cm^2
I_y =	329278	cm^4
$W_{el,y}$ =	9198	cm^3
M_{el} =	2299	kN.m
f_{yk} =	275	MPa
E_a =	210000	MPa

(b)

Characteristic of concrete					
Concrete B1			Concrete B2		
Resistance class		C 50/60	Resistance class		C 25/30
Class of cement		Class N	Class of cement		Class N
A_cB1 =	184000	mm^2	A_c B2 =	576000	mm^2
u B1 =	2060	mm	u B2 =	5280	mm
t_s B1 =	1	jour	t_s B2 =	1	jour
f_{ck} B1 =	50	MPa	f_{ck} B2 =	25	MPa
f_{ctm} B1 =	4,1	MPa	f_{ctm} B2 =	2,6	MPa
E_{cm} B1 =	37000	MPa	E_{cm} B2 =	31000	MPa

Table 1.2 Preflex beam manufacturing steps

Stapes	Tasks	Days
1	Profile preflexion	0
2	Consideration of delay effects of concrete B1	29
3	Relaxation of the preflexion loads	30
4	Consideration of delay effects of concrete B2	59
5	Loading	60
6	Consideration of delay effects at 100 years	36,500

- Step 1. First, we list the lower and upper fiber stresses in concrete B1 at 100 years.
- Step 2. It is then checked whether these values exceed the admissible stress.
- Step 3. If this is the case, we are looking for the date on which concrete B1 is completely cracked. If this is not the case, we go directly to step 6.
- Step 4. Two load cases are created: one before cracking and one after the removal of concrete B1 by adding its own weight.
- Step 5. We relaunch the calculations and we list the stresses suffered by concrete B2 at 100 years because removing concrete B1 increases the stresses in the profile and in concrete B2.
- Step 6. We check that concrete B2 does not reach its admissible limit, and we proceed the same way as for B1 if this is the case. Finally, we list the maximum stress in the steel profile and check that it remains acceptable.

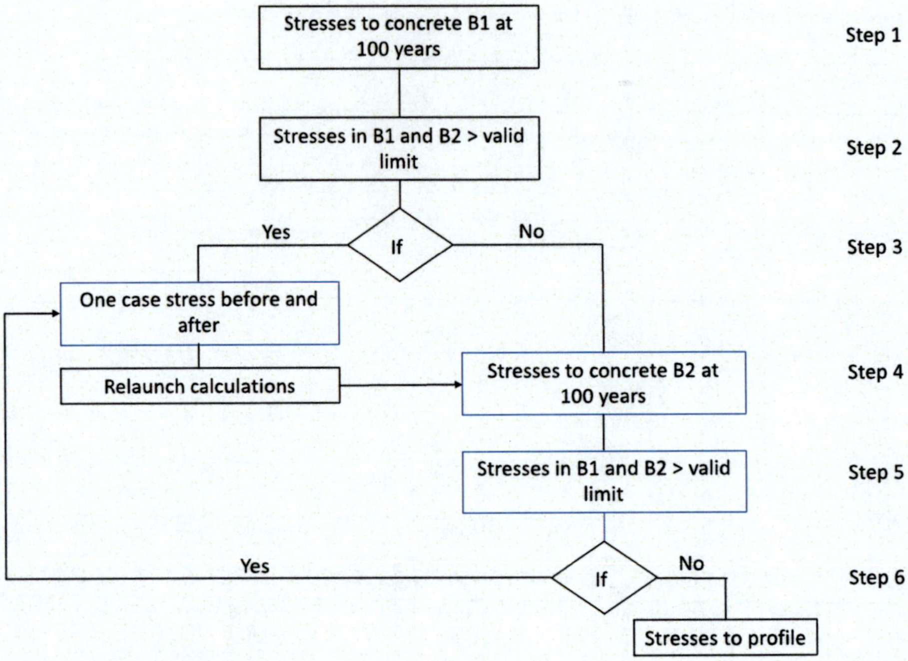

Fig. 1.2 Flowchart for computing the rupture stress

1.3 Results and Discussion

1.3.1 Crack Opening with First Reinforcement

The fracture due to creep and relaxation effects have been discussed in this section. The applying force is 22 kN/m^2 (breaking load of the beam when we do not use reinforcing steel taking up traction in concrete) at the serviceability limit state.

The consideration of cracking of concrete under LUSAS is computed by using steel reinforcements. We therefore started by applying a very weak reinforcement, composed of 6 mm bars every 30 cm, in the longitudinal and transverse direction of the beam, on the upper face and on the lower face.

Figure 1.3 presents the relationship between crack opening versus time for creep and shrinkage effects. It was observed that crack opening due to creep is significantly greater than that due to shrinkage. Indeed, at 100 years, creep represents 77.6% of cracking after loading and shrinkage only 22.4%. On the graph, we can notice that even at 100 years, the cracking induced by shrinkage continues to increase very slightly whereas it is completely stopped for creep.

1.3.2 Crack Opening with Second Reinforcement

In order to decrease the impact of delay deformations on crack opening, the reinforcement has been increased. In this case, bars of 6 mm every 15 cm were placed on the upper side and bars of 12 mm every 15 cm on the lower side. Figure 1.4 presents the relationship between crack opening versus time for creep and shrinkage effects. This reinforcement of steel allows to remain below the admissible limit of cracking, because when we take into account cracking due to loading and delayed effects, we arrive at 0.08 mm at 100 years. Once again, we realize that the cracking due to creep is much greater than that due to shrinkage. The proportion remains the same since 77.6% of the cracking due to delayed effects comes from creep.

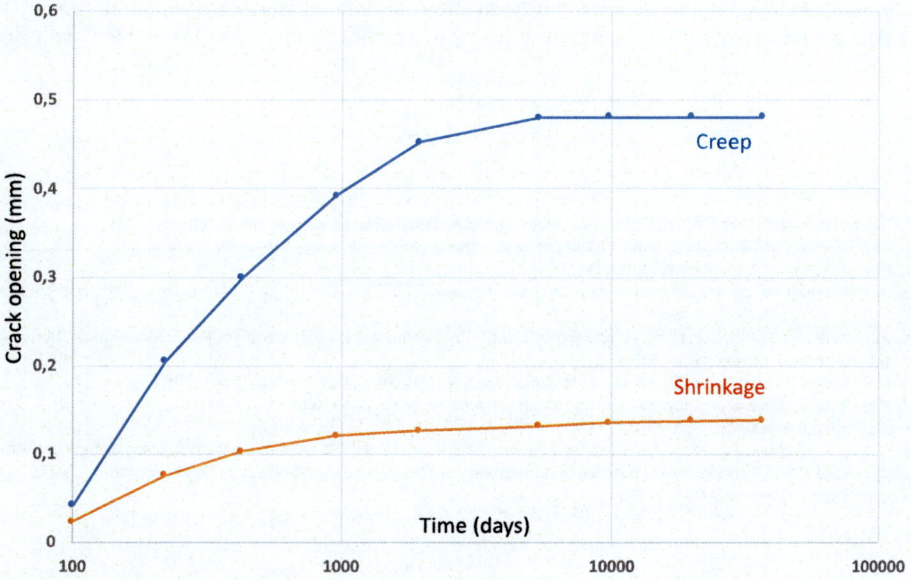

Fig. 1.3 Crack opening due to creep/relaxation effects with first reinforcements

Fig. 1.4 Crack opening due to creep/relaxation with second reinforcements

1.4 Conclusion

In this chapter, the behavior of preflex beam is investigated. The different materials used in the design of this structure are presented. The design step and the numerical implementation of crack opening with LUSAS software are detailed. The crack processes due to creep and shrinkage effects are investigated. It is observed that the use of reinforcement decreases significantly the crack opening even if the creep is responsible for more than 77% of the crack opening process. From these results, it can be concluded that the consideration of delayed effects leads to a lower failure load of the beam. Therefore, it is necessary to consider the delayed effects in the dimensioning of a preflex beam in order to avoid the total collapse of the structure.

References

1. Aribert, J.M.: Construction mixte acier-béton: Calcul des poutres mixtes. Techniques de l'ingénieur. Novembre 2004
2. Aribert, J.M.: Analyse et formulation pratique de l'influence de la nuance de l'acier du profilé sur le degré minimum de connexion partielle d'une poutre mixte. Revue construction métallique n°3 (1997)
3. Bae, D., Lee, K.: Behavior of preflex beam in manufacturing process. KSCE J. Civ. Eng. **8**, 111–115 (2004). https://doi.org/10.1007/BF02829086
4. Foort, E.: Connecteurs dans les poutres Preflex: comportements dus à l'effort normal et à l'effort de rupture – Mémoire de fin d'étude, Polytech Clermont Ferrand. Politechnika Krakowska (2016)
5. Ghali, A., Fabre, R.: Concrete Structures: Stresses and Deformations, 2nd edn. Spon, London/New York (1994)
6. Jerome, M.: Étude de la fissuration des poutres Preflex sollicitées. Comparaions numérique et analytique. Mémoire de fin d'étude, Polytech Clermont Ferrand, Politechnika Krakowska – 2011
7. Zhang, K., Li, S., Liu, K.: Experimental Study on Static and Fatigue Behavior of Steel-Concrete Preflex Prestressed Composite Beam. *Advances in Steel Structures (ICASS '99)*. Proceedings of The Second International Conference on Advances in Steel Structures 15–17 December 1999, Hong Kong, China Volume II, pp. 965–973 (1999)

Chapter 2
Characterization of High Frequency Pulse Loading on Fatigue of Metals

Paul A. Lara, Hugh A. Bruck, and Edda C. Müller

Abstract Aluminum materials of various grades are utilized across many industries, spanning from the cycling, automotive, aerospace, and the marine industry. In the marine grade, aluminum materials are utilized to construct entire vessels of various lengths or portions of them by taking advantage of the lower weight characteristics of the materials and impact on stability of the structures. In particular, 5xxx series aluminum materials are relied on by the marine industry for these purposes, taking advantage of the ability of these series to resist marine corrosive environments. However, during its lifetime, a marine vessel will experience a multitude of variable amplitude loading conditions, with occasional overloads and underloads depending on the operation and environmental conditions. In some cases, these overloads/underloads can result in the failure of the structures by reaching its ultimate capacity, but in other instances, they can systematically affect the growth rate of localized cracks. Existing models, like the Wheeler, Willenborg, plus variations of these, have been utilized to predict the crack growth behavior with varying degrees of success. We created an experimental matrix to explore the effects of overload/underload combinations on fatigue crack growth in 5xxx aluminum. Both visual inspection of crack tip location and digital image correlation (DIC) characterization of the crack tip deformation fields were used to characterize the crack growth in center crack tension (CCT) panel specimens. DIC also enabled additional analysis of strain fields to elucidate on the conditions responsible for change in the crack growth behavior. This chapter outlines some of the ongoing results of this work, which built on past experimental work conducted. Future phases of this work will utilize this data to develop new models for fatigue crack growth, and application of multiple pulses sequences.

Keywords Fatigue crack growth · High frequency pulse · Overloads/underloads · Plastic zone · Crack kinking

2.1 Introduction

Embedded high frequency (HF) signal effects derived from wave impacts on ships can affect failure mechanisms on the structures and have an adverse impact on the fatigue life of the vessel. While operating in a sea environment, ship structures can be subject to many operational loads (wind, pressure, temperature, etc.), one of which is the structural effects [illegible] from the surrounding sea environment. Typically, the wave environment applies an ordinary wave [illegible] drives the primary bending stress of the vessel, along with a more stochastically driven element that [illegible] itself as wave impacts. This dynamic wave impacts results in a high frequency vibratory response sign[illegible] applied to ship structures. The vibratory nature of these high frequency responses makes the design of structures increasingly complex in nature due to their uncertainty; designers and naval design rule societies have relied on methods such as safety factors and/or margins of safety to account for its effects. A typical wave impact on a naval structure imparts a time-dependent pulse load that exhibits a higher frequency logarithmic decaying sinusoidal response when measured via experimental means and captured utilizing

P. A. Lara
Naval Surface Warfare Center, Carderock Division, West Bethesda, MD, USA
e-mail: lara@umd.edu

H. A. Bruck (✉)
Mechanical Engineering Department, University of Maryland, College Park, MD, USA
e-mail: bruck@eng.umd.edu

E. C. Müller
Düsseldorf University of Applied Science, Düsseldorf, Germany
e-mail: edda.mueller@study.hs-duesseldorf.de

S. Xia et al. (eds.), *Fracture, Fatigue, Failure and Damage Evolution, Volume 3*, Conference Proceedings of the Society for Experimental Mechanics Series, https://doi.org/10.1007/978-3-030-60959-7_2

data acquisition systems. The figure below displays one of these signals, where the measured signal can be decomposed into its different components by applying filtering techniques.

Under the current practices, the peaks of these responses are utilized to compare to the quasi-static design stresses as well as evaluate the fatigue life of the structures either under constant or variable amplitude experimentation. Figure 2.1 below displays this kind of behavior, where underloads and underloads affect the crack growth rate behavior of material under a simplified signal approach [1].

But there are additional higher frequency characteristics to these high frequency signals that are not currently taken into consideration. Existing academic research has been centered on capturing a simplified sinusoidal response associated with this slamming event and embedded high frequency response but has not addressed logarithmic decay, signal frequency, or frequency of occurrence. Work by Sumi et al. has centered on more complex signals, where bounding functions were used to predict crack growth behavior [2], and the results show an influence of crack length predictions based on utilizing these bounding functions.

All these factors have associated uncertainty and impact fatigue life and failure mechanisms exhibited by ship structures; work by researchers such as Fricke et al. has shown damage variation related to the amplitude of high frequency signals [3]. These high frequency effects can cause retardation of crack growth, and foundational models such as Wheler [4] and Willenborg [5] were developed to capture the overall effects of these overloads and underloads. Follow-on work was carried out to gain better understanding of the conditions present, as shown in Fig. 2.2, where the crack growth effects within the overload plastic zone boundary exhibits a variable crack growth rate behavior instead of constant crack growth retardation effect.

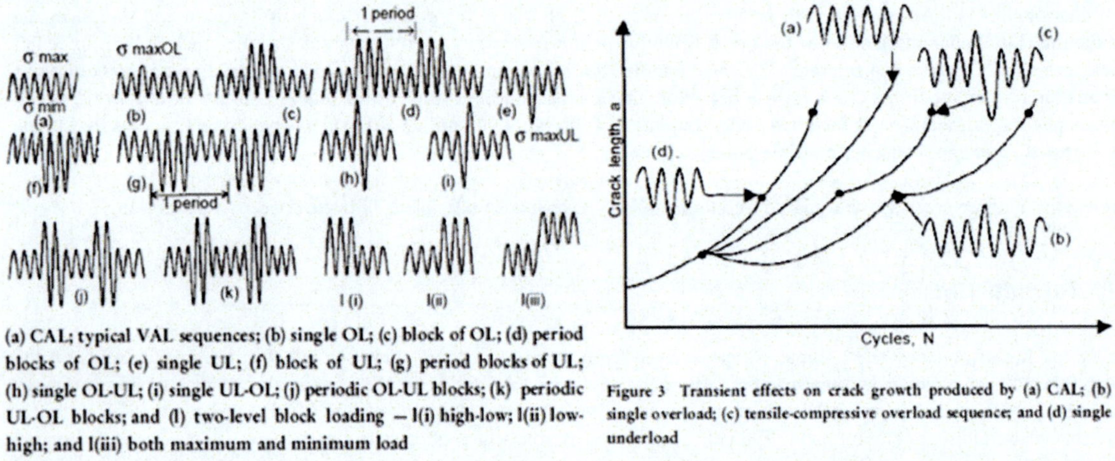

Fig. 2.1 Load sequencing [2]

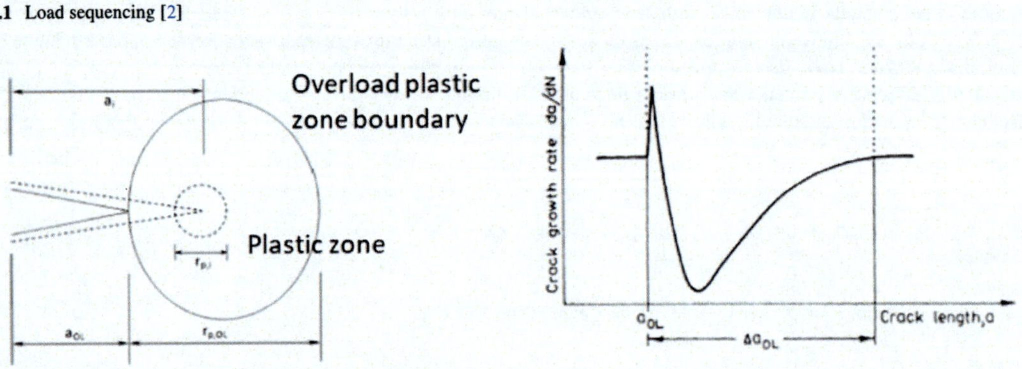

Fig. 2.2 Plastic zone model and retardation effects, left [6], right [7]

Existing literature gaps are: a deeper understanding of resulting fundamental inelastic aluminum behavior for center crack tension (CCT) specimens due to HF pulses elucidate the plastic zone behavior and resulting material failure modes by leveraging full strain field displacement measurements from digital image correlation (DIC) and develop failure models that accounts for observed material failure behavior due to HF pulses and overall fatigue life response. In the future, the foundational findings of this research can then be leveraged in investigative efforts to develop analytical models addressing heat-affected zones, provide underpinnings for high fidelity numerical modelling, eventually providing broader toolsets to designers that can be used to reduce the dependency on safety factors and introduce more rigorous failure mechanism design criterion.

The work outlined here builds on previous work carried out by the authors; it summarizes some previous work, presents new findings, highlights future efforts, and in the end has an overall objective to gather fundamental understanding of the effects of this high frequency loading on Aluminum 5xxx material, accounting for some of the signal's characteristics, and through an experimental evaluation assesses its impact on the local failure mechanism and life cycle models.

2.2 Description of Experiments

This section outlines a summary of the experimental setup; the work builds from and complements existing work developed by Hart and Bruck [8], where the investigators looked at failure of aluminum 5052-H32, the applications of composite patches to aluminum, and the effects of crack growth. Whereas their work centered on the utilization of a composite laminate to mitigate crack growth on ultimate failure, this work looks at the effects of crack growth based on loading condition. Figure 2.3 below displays the aluminum material properties that Hart and Bruck obtained utilizing ATSM 8 [9] and American Society for Testing and Materials (ASTM) E466 [10], as well as the layout of their center crack panel specimens. The Modulus listed for the material is 70.9 MPa, the yield at 2% offset is 116 MPa, and the parameters α and n at 9.9 and 0.22, respectively.

Initial experiments were carried out on this setup for this work, utilizing a two-post Bose ElectroForce frame which has the capacity of applying loadings up to 100 Hz and 3000 N. This device has a built-in displacement meter with errors less than 5 μm, and it has a block waveform generation software that allows for the implementation of unique load signals. Fast Fourier Transforms methods were utilized to ensure that the programmed signal and the applied signals displayed the required frequencies. But the initial results indicated that there was an imbalance of crack growth between the left and right cracks growing from the notch during fatigue loading, the reason as traced to the use of mechanical wedge friction grips. It is important to mention that the experiments by Hart and Bruck utilized a 4-post MTS servo hydraulic grip setup

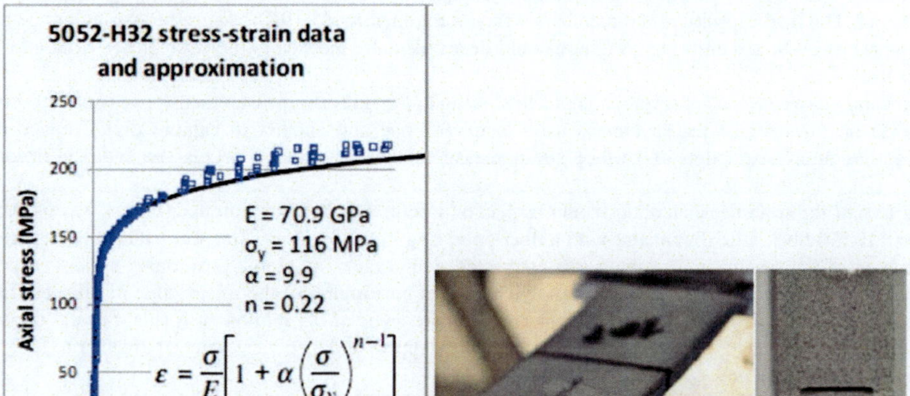

Fig. 2.3 5052-H32 material properties and specimen failure [5]

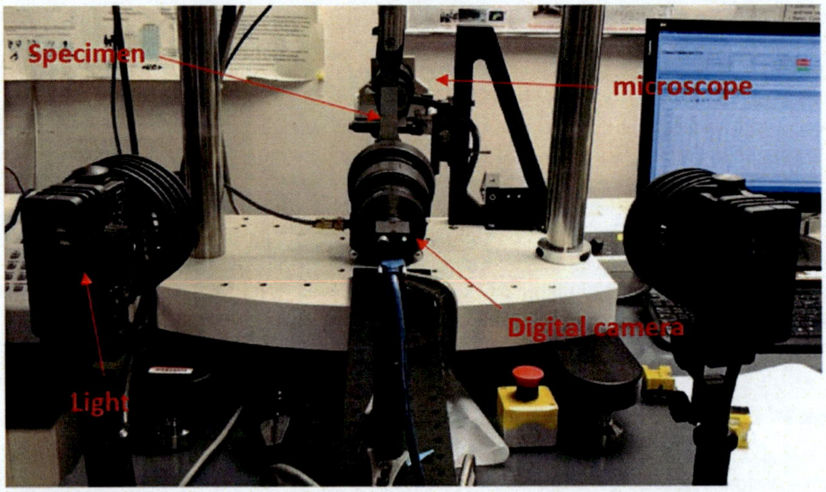

Fig. 2.4 Experiment setup

since they were interested in the ultimate loading capacity. To address this crack asymmetry from loading imbalance, a specimen configuration was developed, and it is an amalgamation of several ASTM methods; this was conducted to address the significant asymmetry observed with crack growth, as previously described. ASTM E8 [9] guidance was utilized to develop the specimen tapers toward test region, while ASTM E647 [11] middle tension specimen configuration was utilized to size the center notch and test region, maintaining the configuration carried out by Hart and Bruck. Since a clevis pin was utilized, ASTM E338 [12] was utilized to develop end reinforcements and alignment tabs for the end pin connections to ensure the failure would not occur in or adjacent to the clevis pin area.

Electrical discharge machining (EDM) techniques were utilized to manufacture the specimens, providing a high level of accuracy for the component geometry, since it was also critical to ensure alignment of end pins and size and location of center notch. The material for the specimen is marine grade aluminum 5052-H32, with a nominal thickness of 2.29 mm (0.09in). Figure 2.4 shows how the clevis pins were utilized in the loading train and end adaptors were used to interface with the Bose frame and to help with specimen load alignment; the Universal Grip LLC. components were utilized for this setup: Adapter - Male Bolt 5/8″–18 w/1″ length, Universal Joint Adapter; in addition, Mark-10 G1090 high-capacity clevis grip (pair) were utilized. The load capacity of the members was at a minimal load of 9KN, hence far exceeding the capacity of the Bose equipment of 3KN, and ensuring no failure could be accidentally induced to the Bose fixture from a failure of the adapters under load.

The experimental setup consisted of several components' setup to provide dual measurements: one was to observe/capture the crack growth on the back of the specimens using a microscope and another to capture digital image correlation information for the future extraction of surface strain measurements. Figure 2.4 details the setup utilized for these experiments.

In the back face of the specimens, an optics zoom inspection microscope head by Edmunds Optics was used along with a Dolan-Jenner MI-150 fiber optic illuminator with a fiber optic ring light accessory. This was utilized to capture the crack growth on the back of the specimen. The setup was connected with a rack and pinion positioning system, which allowed movement of the microscope along three axes to allow consistent positioning of the microscope in different locations to capture crack growth. The microscope was able to visualize square areas of 0.5 mm on each side. Several digital pictures were taken on each crack emanating from each notch, which overlapped, and then were stitched together to form an overall view of the crack of the specimen.

In the front face of the specimen, a Point Gray Flea FL3-U3-88S2C-C digital camera attached to a Tamron 28–80 mm f/3.5–22 lens and an adapter was used to capture static pictures of the surface of the specimen. In addition, two 250 Watts halogen Lowel Pro Lights were utilized to provide lighting to the specimen. The lights set on flood (wide) mode were not directly aimed at the specimens, but instead aimed at locations adjacent to the mid-point of the specimen, and focused as to provide evenly dispersed light flooding on the specimen. Digital image correlation (DIC) was utilized on the front face of the specimen to capture field surface strains [13]; in order to capture these images, the area was prepped to generate distinct patterns that can be captured with a Point Grey FlyCap 2 V2.11.3 and processed utilizing VIC-2D Ver6.0.6 by Correlated Solutions (Lexington, SC).

Figure 2.5 shows the typical DIC pattern applied; the generation of this pattern is achieved by first applying a coat of flat white paint as the background, and the black speckle patterns rely on lightweight particles that slowly settle over the specimen. Best practices developed throughout several references were utilized to obtain acceptable DIC results, once such body of work was developed by the international digital image correlation society [14].

2.3 Test Matrix for HF Pulse Loading Conditions

In order to explore the variability of the high frequency pulses, a couple of items were developed: the first was a consistent process/protocol that was repeatable across experiments, and the second was an experimental matrix that varied certain parameters to determine their influence on the response. Figure 2.6 below outlines the top-level experimental process utilized.

Specimen preparation was the initial step in the configuration; the back of the specimens that faced the microscope was polished to a mirror-type finish; this aided in the identification of the crack tip. The front of the specimens received a DIC spackle pattern that had a sigma level equal or below 0.001 [13]. The sigma level was tested during the baseline measurements stages until the desired value was achieved. The baseline measurements also utilized DIC to gather initial measurement errors bounds; if the load frame indicated a 0.2 mm displacement, did the DIC measurements also show a similar value? This was done by only pining the stationary end of the specimen and comparing the displacement between the load frame and DIC calculations. Additional baseline measurements included static DIC measurements at 0, 200, 1100, and 2000 N, as well as microscope crack measurements at no load and 2000 N. The initial crack growth was a step used to develop a crack and go outside the envelope of the stress-strain effects of the notch radius [15]. The notch radius was 0.17 mm, and an initial starter crack was developed to be around 1 mm in length. The HF pulse was applied at about 40 K cycles when the desired crack length was reached; subsequent measurements monitored the crack growth rates after the HF pulse was applied to evaluate the recovery. It is important to note that the load ranged from 200 N to 2000 N, where the stress ratio $R = 0.1$, the frequency used for fatigue was 5 Hz, and the nominal stress intensity $\Delta K = 6.07$ MPa-m$^{0.5}$.

The experimental matrix utilized is shown in Table 2.1; it consists of a baseline test where no HF pulse is applied, and a 3×3 matrix with three levels and two factors (amplitude, frequency), resulting in 9 treatments, where the amplitude of the pulse is increased in 1/3 amplitude increments and frequencies that range between 4 and 10 times the baseline frequency, i.e., specimen 3C had a peak amplitude of 2900 N and an HF pulse with a frequency of 50 Hz.

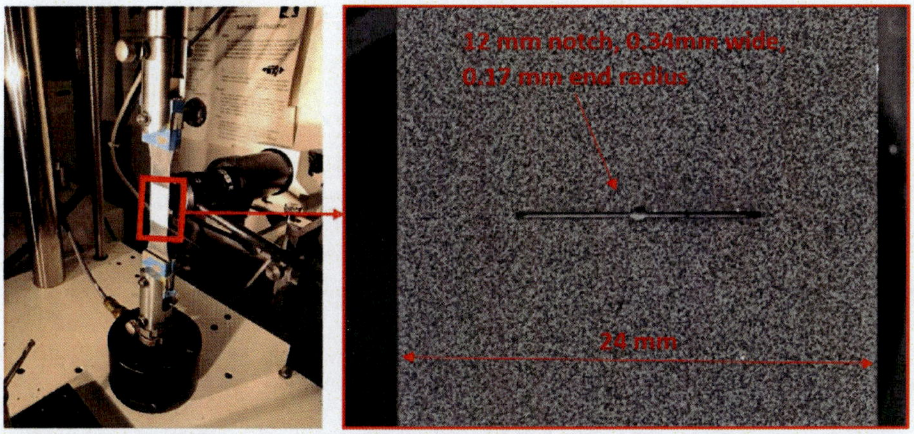

Fig. 2.5 Specimen and speckle pattern

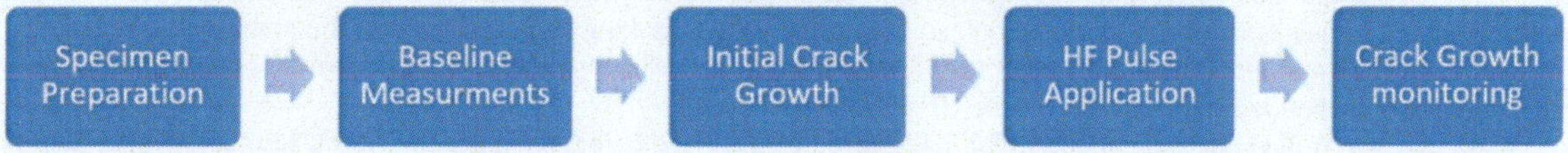

Fig. 2.6 Experimental process

Table 2.1 Test matrix for HF pulse loading conditions

		High Frequency Pulse Amplitude (N)		
	Baseline, No Effects	1/3 Peak Amplitude 2300	2/3 Peak Amplitude 2600	Peak Amplitude 2900
High Frequency Pulse Frequency (Hz)	4 x Main Wave 20	1A	1B	1C
	7 x Main Wave 35	2A	2B	2C
	10 x Main Wave 50	3A	3B	3C

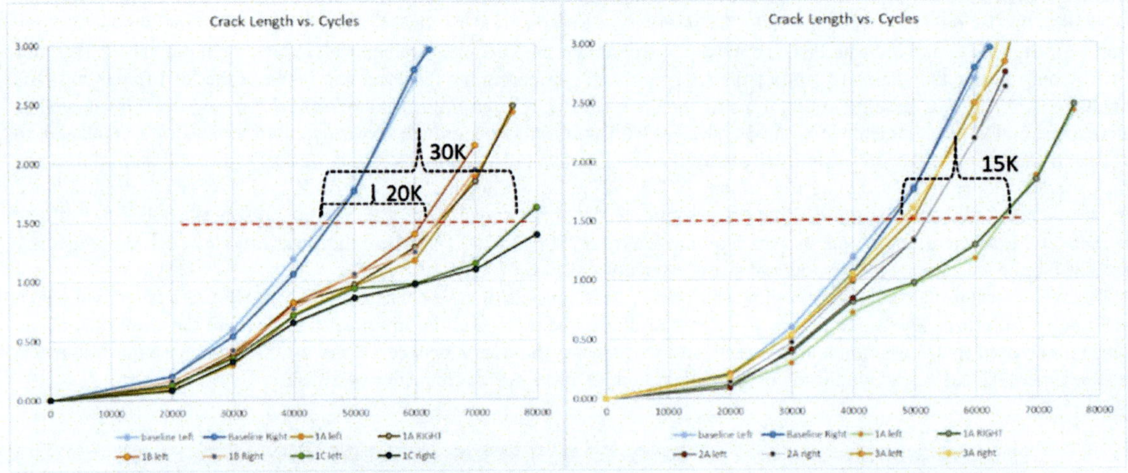

Fig. 2.7 Crack length for varying load (left) and frequency (Right)

For this matrix, the HF pulse was only applied at the peak of the ordinary wave of 5 Hz; hence the phase shift was held constant as well as the log decrement of the function was maintained at 0.21. Introduction of other factors would increase the matrix size to a 3^3 design resulting in 27 treatments and are recommended for future experiments.

2.4 Experimental Results

When plotting the data, as shown in Fig. 2.7, the initial results were the crack length for specimens where the load is varied while the frequency is held constant (left) and specimens where the frequency is held constant and the load is varied (right). One can see that when comparing the number of cycles it takes to grow a crack to a nominal length of 1.5 mm, the varying load (left) has a noticeable effect on the crack growth. It takes 20–30 K cycles in delay depending on the load. One can see that the number of cycles it takes for the crack to grow 0.5 mm on the baseline is substantially smaller that the cycles needed for the specimens that had an HF pulse applied. When looking at the varying frequency (right), one can see that a lower frequency has a higher effect for the same max load of 2300 N. When looking diagonally across the matrix, one can see in Fig. 2.7 that the lowest block with the lowest load and frequency behaves very similarly to the baseline, but other blocks begin to deviate as the load and frequency of the HF pulse are increased.

Vic2D DIC was utilized to evaluate the strain fields ahead of the crack before and after the HF pulse was applied. Figure 2.8 shows the various stages of the strain changes surrounding the crack; in addition, the complete crack path is also shown in relation to the strain zones. At 40 K cycles, the crack is moving in a downward direction, and upon application of the HF

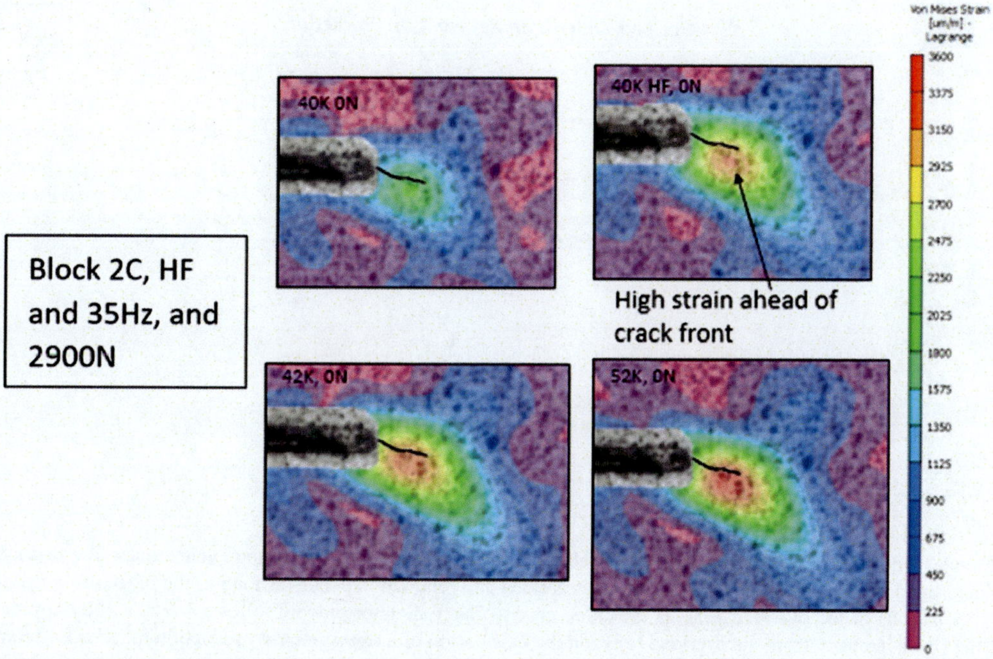

Fig. 2.8 Strain field evolution, block 2C at HF of 35 Hz, and peak amplitude of 2600 N

Wheeler

$$\frac{da}{dN} = \Phi_R \left(\frac{da}{dN}\right)_{CAL}$$

$$\text{where} : \Phi_R = \begin{cases} \left[\frac{r_{p,i}}{a_{OL} + r_{p,OL} - a_i}\right]^m & \text{when} \quad a_i + r_{p,i} < a_{OL} + r_{p,OL} \\ 1 & \text{when} \quad a_i + r_{p,i} > a_{OL} + r_{p,OL} \end{cases}$$

Willenborg

$$\frac{da}{dN_i} = \frac{C(\Delta K_{eff})_i^m}{[(1 - R_{eff})_i]K_c - (\Delta K_{eff})_i}$$

Fig. 2.9 Wheeler and Willenborg models

pulse, a high strain hardening region is developed ahead of the crack, acting like a hard inclusion. The DIC data show how strains changed from 42 K cycles to 52 K cycles, including its size and dimensions; it includes utilization of work by Bruck [16], and modification work by Yates was utilized to locate the crack tip location [17] and associated plastic zone [18, 19]. The DIC data capture a kinking of the crack, also observed in microscope measurements, and how it circumvents around this are of high strain in order to continue its growth. This highlighted future work needed in investigating the strain fields ahead of the crack and shear effects to better understand the kinking behavior. This all leads to gaining a deeper understanding of the behavior observed by Laseure [7].

At the core of this work are the models by heeler [4] and Willenborg [5]. These models look at the plastic zone behavior, Wheeler relies on crack growth rates, and Willenborg utilizes stress intensity for the calculation, as shown in (Fig. 2.9):

More recent research work by Mehrzadi and Taheri [6] and Chen [19] outline the behavior of the plastic zone ahead of the crack tip. The common thread of all these models is that an overload plastic zone is created once an overload/underload occurs, and the magnitude of this region has an effect on the crack growth rate as long as the cyclic plastic zone stays within its boundaries. The radii of the plastic and overload zone can be calculated utilizing the Irwin approximation [20]. The

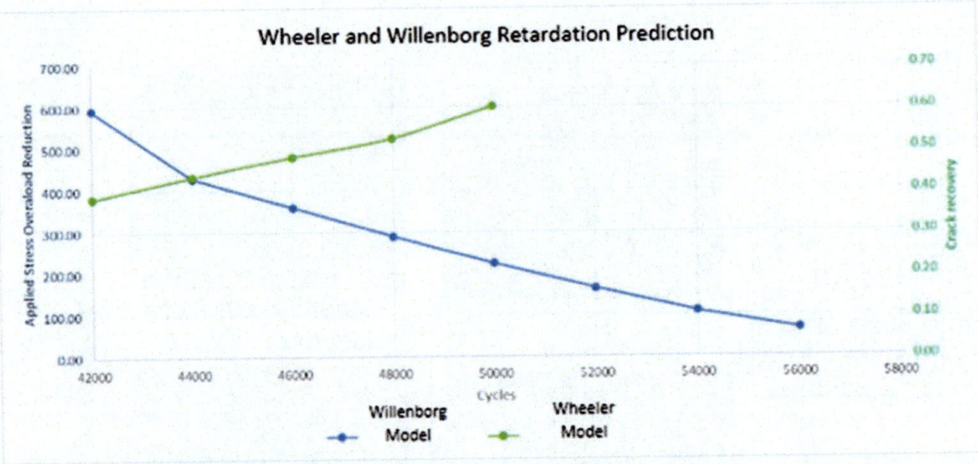

Fig. 2.10 Wheeler and Willenborg retardation predictions of HF pulse of 50 Hz and 2900 N

Wheeler model relies on the Irwin calculation to develop the radii of the overload zone and cyclic zones. The methodology of the Wheler model is that an overload plastic zone is created and retardation occurs as long as the reoccurring cyclic zone is within the bounds of the overload zone as the crack grows. The Willenborg model follows a very similar approach but relies on the stress intensity from the overload [15] and the calculation of a stress intensity reduction in order for retardation effects to seize. This behavior can be seen graphically showing how the residual stress intensity is accumulated after the HF pulse at 42 K cycles and how it diminishes over time to no retardation at 56 K cycles for the Willenborg model, Fig. 2.10. For the Wheeler model, the crack recovery can also be plotted to show when the radii of the cyclic plastic zone reaches the boundary of the overload zone and also hence ends the retardation cycle at 50 K cycles.

One of the novel findings of this work was the observation of a crack kinking mechanism resulting from the application of certain HF pulses, outlined in the next section.

2.5 Crack Kinking Behavior

An observed crack kinking behavior was mentioned in previous sections, during the experiments; this behavior was observed and tracked for the various specimen conditions outlined in Table 2.1. The microscope data shown in Fig. 2.11 show the crack growing from the left notch of each specimen, the orange arrow marks the direction of crack growth, the red mark shows the application of the applicable HF pulse, the green marks delimitate locations of data extraction in the fatigue cycle.

While the microscope measurements provide a view of the crack propagation, a Barker's etch method [21] was used for anodizing the aluminum alloy and then utilizing optical microscopy with polarized lights to observe the grain boundary and the propagated crack. This was evaluated at 500 μm, 100 μm, and 20 μm magnification levels, allowing the evaluation of the crack propagation through the grain, evaluating if there was a presence of large material inclusions, as well as investigating shielding effects such as contact and deflection on crack growth [22]. Fractography was also utilized to examine the surfaces through the crack depth on the specimen; several test specimens were carefully cut to allow views of the through thickness crack front. The results showed a distinct beach mark shape due to the HF pulse and a change in contrast near the edge surfaces due to the inclination associated with crack kinking. This evaluation also showed that the width of the beach mark is consistent with plastic zone and narrows beneath the surface as it is typical of plane stress to plane strain transition regions. This assessment did not indicate that there was a presence of a material inclusion that could cause the observed kinking behavior.

The crack kinking behavior observed can be segmented into several areas associated with the kinking condition, as shown in Fig. 2.12. After the application of the HF pulse, there is a section when the crack continues to grow, and before it kinks, another section is when the crack is fully kinked and continues in that condition, and a last section where the crack kinking behavior has ceased and the crack has returned to its semi-original growth path. Length and angle parameters were gathered for the crack.L1 defines the crack length after the HF pulse and before the kink, L2 defines the length of the kink. The angle

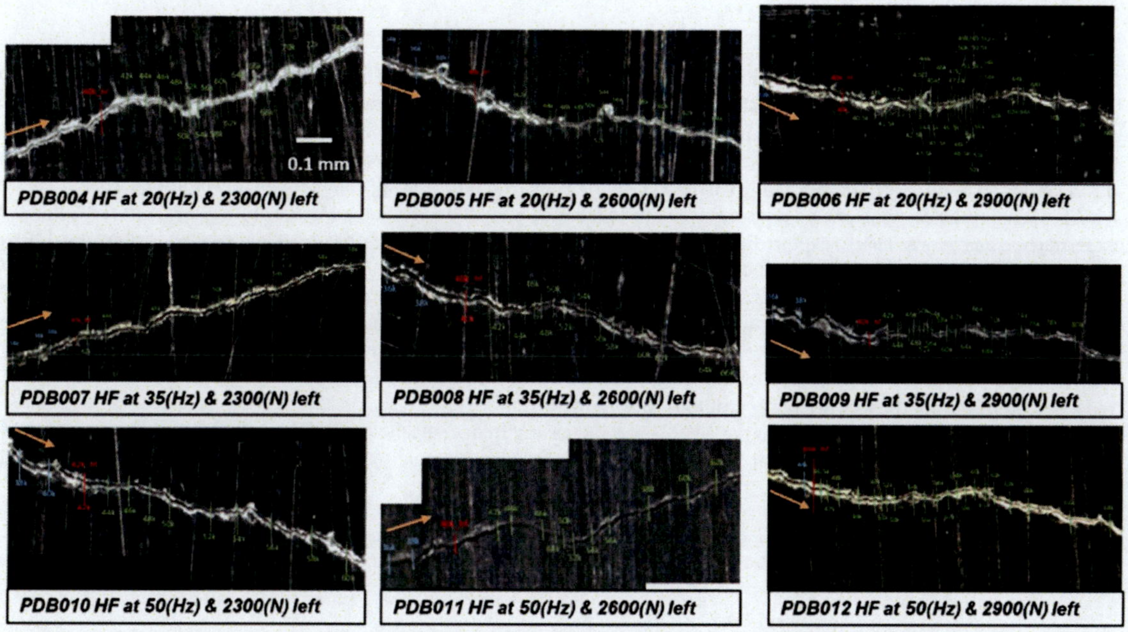

Fig. 2.11 Crack kinking behavior matrix

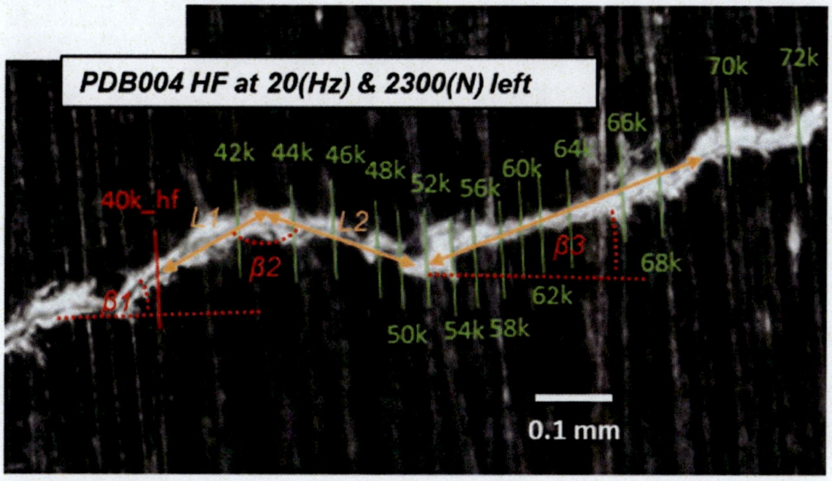

Fig. 2.12 Crack kinking characterization

β1 defined the entry angle in relation to the horizontal plane, β2 is the kink angle between L1 and L2, and β3 is the exit angle with respect to the horizontal.

It is worth noting that the entry angle β1 is dependent on the original direction of crack growth and independent of the HF pulse condition, as specimens do not have a consistent entry angle β1 of crack growth, which can be dependent on notch conditions and presence of inclusions during initial crack growth. The crack kink data highlighted several findings; one was that the kink angle β2 is consistent between experiments and has a mean of 136 degrees as measured between L1 and L2. Another finding was that the exit angle β3 is consistently less than the entry angle β1 by a mean of 2 degrees.

2.6 Effects of HF Pulse on Da/DN

The data also allowed for the identification of the variation of Da/DN, as shown in Fig. 2.2, but with higher fidelity to who show the effects of an application of the HF pulse. In Fig. 2.13, the left figure shows that retardation effects are quite consistent over several test specimens that received the same HF pulse, but these effects are not quite present as shown on the right figure, where the HF pulse increased in frequency to 50 Hz, but maintained a maximum load of 2300 N. This observation can also be seen on Fig. 2.11, where no kinking is observed for the specimen on the lower left quadrant.

While the results do not show a very clear crack retardation from certain HF pulses, it does provide data to show that the experimental matrix was able to capture limitations toward the presence of kinking behavior, where certain loading conditions do not precipitate a clear kinking of the crack, but instead less visible retardation effect. By leveraging the characterization of the crack kinking, as shown in Fig. 2.12, it is possible to examine the crack growth rate for varying frequency and load. Figure 2.14 shows these values and certain observations can be gathered from the results. The behavior of increased crack growth rate after the HF pulse and decrease during the kink portion can be observed. Another observation is that by varying the HF pulse form 2300–2900 N and maintaining a frequency of 20 Hz, affects the Da/DN after the HF pulse, but also reduces Da/DN during the kinking portion of the crack growth.

However, by holding the load constant at 2900 N and increasing frequency, the Da/DN during the kink actually increases. This data was very promising as it shows The DIC data allows for the interrogation of surface strains ahead of the crack path

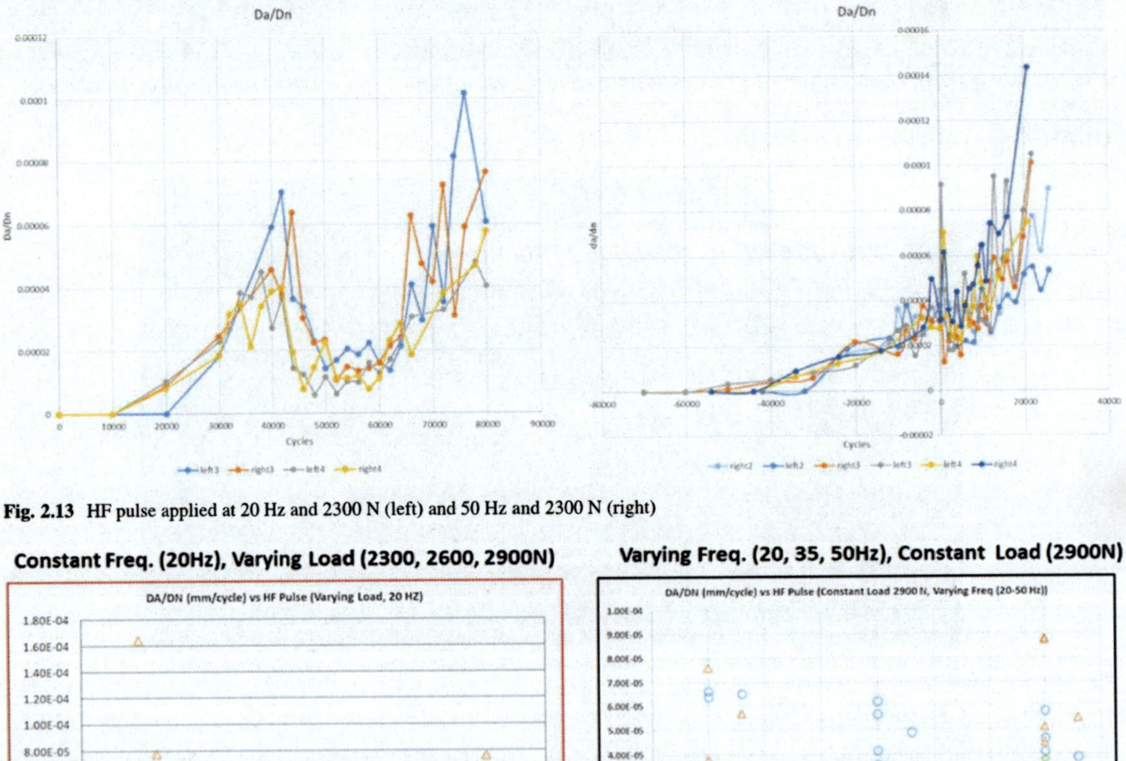

Fig. 2.13 HF pulse applied at 20 Hz and 2300 N (left) and 50 Hz and 2300 N (right)

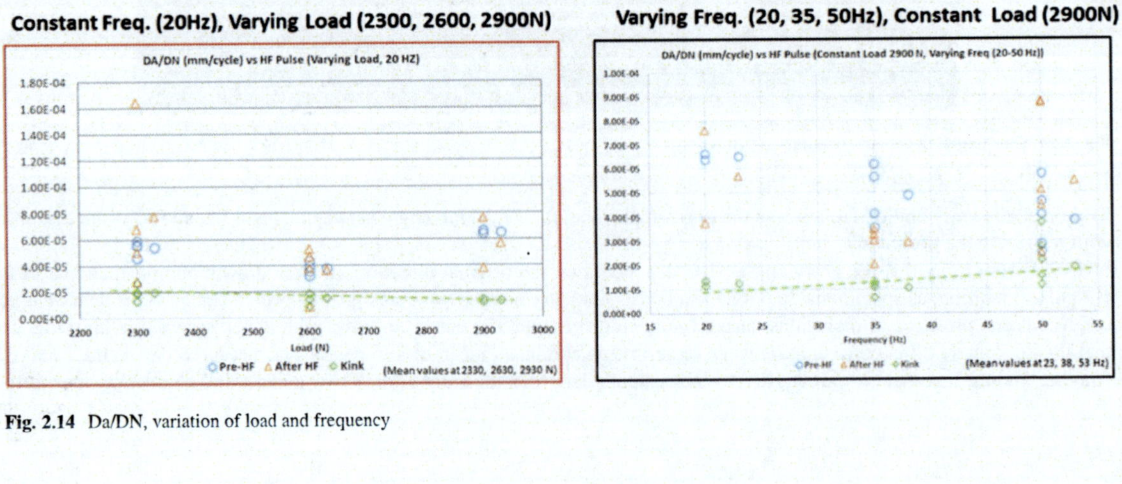

Fig. 2.14 Da/DN, variation of load and frequency

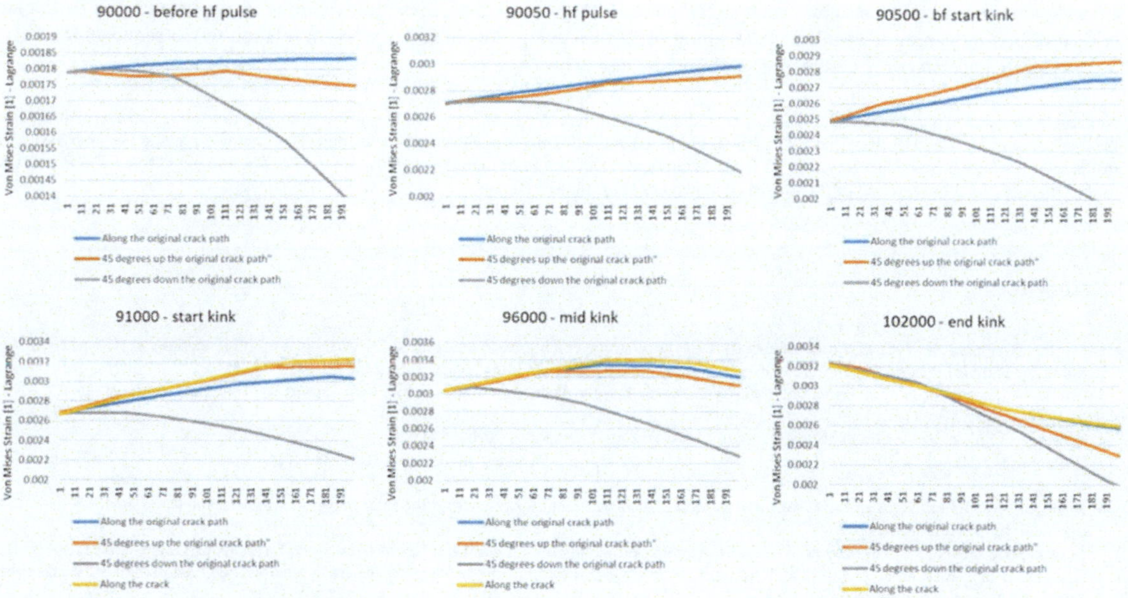

Fig. 2.15 Von Mises strain fields ahead of crack

along different vectors, and to see how they evolve over time. Figure 2.15 shows these Von Mises strain fields at ±45° from the crack direction, as well as along original crack path. That the matrix captures a variation of crack growth rates, as well as kinking behavior which is affected by both magnitude and frequency.

The fanning of the strain fields changes due to the application of the HF pulses and has a very distinctive variability when the kink is complete.

Work is ongoing to utilize DIC strain and crack opening displacement (COD) to develop a flow model that steps through different conditions that would trigger kinking effects and their termination. As initially expected, DIC data are providing critical data in understanding how the surface strains vary and their correlation to the effects of HF pulses.

2.7 Conclusions

Existing models like the Wheeler, Willenborg, and variations of these have been utilized to predict the crack growth behavior with varying degrees of success. In this work, we carried through an experimental matrix to explore the effects of HF pulses on fatigue crack growth in 5xxx aluminum, typically used in marine applications. We utilized visual inspection of crack tip location and used digital image correlation (DIC) techniques for characterization of the crack tip deformation fields. These techniques were used on a designed center crack tension (CCT) panel specimens to measure crack growth. DIC techniques also enabled additional analysis of strain fields to better understand the associated crack growth behavior. A novel crack kinking behavior was observed from certain HF pulses; this kinking behavior was characterized and several metrics were extracted to show its influence on crack growth rates and retardation. This ongoing work includes utilizing this observed data to finalize an embedded high frequency load models for fatigue crack growth and exploring the phases of HF pulses.

2.8 Future Work

This experimental data is providing basin information toward the kinking behavior and crack retardation from the application of HF pulses, but it also highlights areas where additional work is needed. There are long-term efforts and short-term actions on this. Long term there is the evaluation of heat-affected zones and weld effects, introduction of other materials, and expansion to include other variables such as HF pulse shift or decay rates. Short-term actions aim to target some particular critical efforts in gaining a deeper understanding of this behavior. This includes work in relation to the sequencing of HF pulses to gain understanding on the effects of kinking and delay.

Acknowledgments Work described was performed by the Naval Surface Warfare Center Carderock Division's Platform Integrity Department and the University of Maryland College Park's Department of Mechanical Engineering. Financial and technical support was provided by an NSWCCD In-house Laboratory Independent Research (ILIR) program under Dr. Jack Price and a grant provided to UMD by Program Officer Dr. Paul Hess of the Office of Naval Research Code 331 under grant number N000141812016.

References

1. Murthy, R., Palani, G., Iyer, N.: State-of-the-art review on fatigue crack growth analysis under variable amplitude loading. IE(I) J., 12 (2004)
2. Sumi, Y.: Fatigue crack propagation in marine structures under seaway loading. Int. J. Fatigue, 7 (2014)
3. Fricke, W., Paetzold, H.: Experimental Investigations on Fatigue Damage of Ship Structures Caused by Whipping Stresses. In: PRADS, Changwon City, Korea (2013)
4. Wheeler, O.: Spectrum loading and crack growth. J. Basic Eng. Transport. ASME, 5 (1972)
5. Willenborg, J., Engle, R., Wood, H.: A crack growth retardation model using an effective stress concept. AFFDL TM-71-1-FBR, Jan 1971 (1971)
6. Mehrzadi, M., Taheri, F.: A material sensitive modified wheeler model for predicting the retardation in fatigue response of AM60B due to an overload. Int. J. Fatigue, 10 (2013)
7. Laseure, N.S.I.M.N.D.W.W.: Effects of Variable Amplitude Loading on Fatigue Life. Ghent University, Labo Soete, Ghent (2016)
8. Hart, D., Bruck, H.: Characterization and modeling of low Modulus composite patched center crack tension specimen using DIC surface. In: Society of Experimental Mechanics (2018)
9. ASTM-E8: ASTM E8M-08: Standard Test Methods for Tension Testing of Metallic Materials. ASTM International, Conshohocken (2008)
10. ASTM-E466: ASTM E466–07: Standard Practice for Conducting Force Controlled Constant Amplitude Axial Fatigue Tests of Metallic Materials. ASTM International, Conshohocken (2007)
11. ASTM-E647: Standard Test Method for Measuring Fatigue Crack Growth Rates. ASTM International, Conshohocken (2015)
12. ASTM-E338: Standard Test Method of Sharp-Notch Tension Testing of High-Strength Sheet Materials. ASTM International, Conshohocken (2003)
13. Sutton, M., Orteu, J.-J., Schreier, H.: Image Correlation for Shape, Motion, and Deformation Measurements. Springer LLC, Boston (2009)
14. Jones, E.: Good practices guide for digital image Correllation. Int. Digital Image Correl. Soc. (2018)
15. Bannantine, J., Comer, J., Handrock, J.: Fundamentals of Metal Fatigue Analysis. Prentice Hall Inc, Upper Saddle River (1990)
16. Bruck, H.: Analysis of 3-D Effects near the Crack Tip on Rice's 2-D J-Integral Using Digital Image Correllation and Smoothing Techniques, M.S. Thesis, Univeristy of South Carolina (1989)
17. Yates, J., Zanganeh, Y.: Quantifying crack tip displacement fields with DIC. Eng. Fracture Mech., 14 (2010)
18. Gdoutos, E.: Fracture Mechanics, An Introduction, 2nd edn. Springer, Minneapolis (2005)
19. Chen, F., Wang, F., Cui, W.: Fatigue life prediction of engineering structures subjected to variable amplitude loading using the improved crack growth rate model. Fatigue Fracture Eng. Mater. Struct., 13 (2011)
20. Anderson, T.: Fracture Mechanics Fundamental and Applications, 3rd edn. Taylor and Francis Group LLC, Boca Raton (2005)
21. Cerri, E., Evangelista, E.: Metallography of Aluminum Alloys. Mechanical Engineering Department, European Aluminum Association, Ancona (1999)
22. Ritchie, R.: Mechanisms of fatigue crack propagation in metals, ceramics and composites: role of crack tip shielding. Mater. Sci. Eng. **A103**, 14 (1988)

Chapter 3
Fracture Parameters and Failure Visualization of Al6063-T6 Under Different Loading Rates

Anoop Kumar Pandouria, Purnashis Chakraborty, Sanjay Kumar, and Vikrant Tiwari

Abstract Dynamic fracture analysis of aluminum alloys is a significant research area due to their uses in the aerospace and defense industry, where catastrophic failure can occur suddenly. The crack mouth opening displacement (CMOD) is an important tool in determining the various fracture parameters like stress intensity factor. Quasi-static and dynamic three-point bend (TPB) experiments were performed to investigate the effect of high loading rates on CMOD and fracture behavior of the Aluminum 6063-T6 alloy. A 3D Digital Image Correlation (DIC) was utilized in measuring the CMOD experiments. Quasi static three-point bend experiments were performed using UTM and high-resolution stereo camera system for low loading rates. For dynamic three-point bend experiment, a modified Hopkinson pressure bar setup, two high speed digital cameras, and a dedicated data acquisition system were utilized in measuring load point data and the full field displacement profile. Authors have also evaluated the time response of static and dynamic stress intensity factor from the CMOD with the help of the standard formulations as used by several other researchers. Authors found the values of dynamic fracture toughness are significantly high as compared to static fracture toughness.

Keywords Dynamic three-point bend experiments · CMOD · 3D stereoscopic DIC

3.1 Introduction

One of the desired quality for lightweight vehicles is to resist the fracture during the crash. In recent days, automotive manufacturers have used different aluminum alloys for lightweight vehicles to improve fuel efficiency and increase vehicles' crashworthiness. During the impact on any mechanical systems, the damage mode, the stress state, and fracture mechanisms are different at different locations. Thus, the knowledge of fracture behavior under high strain rate is important to understand the response of structural components subjected to dynamic loading.

Fracture toughness is the most important parameter to define the fracture criteria for the component which commonly is the value of the stress intensity factor (SIF) at the time of crack initiation. Various methods are available for calculating static stress intensity factor, but no standard methodologies have yet been established for the dynamic case. The instrumented Charpy test is widely used for dynamic failure analysis; different researchers [1–4] have also used a split Hopkinson pressure bar for high strain rate characterization. Ming-zhi XING et al. [5] have investigated the dynamic fracture initiation toughness of a notched three-point bend (TPB) specimen and also analyzed the fracture surface. Both Static and dynamic SIF (DSIF) can be calculated with the help of crack mouth opening displacement (CMOD). Several noncontact experimental techniques are currently available that utilize the light intensity to measure the in-plane and out-of-plane displacements, but the digital image correlation (DIC) [6] is most commonly used. In this chapter, we use 3D DIC technique to measure the CMOD and crack initiation for both the cases. The quasi-static three-point bend (TPB) test was performed using a universal testing machine, coupled with high-resolution cameras, according to ASTM standard procedure [7]. Similarly, the modified Hopkinson pressure bar (MHPB) setup coupled with high-speed cameras was used to perform a dynamic three-point bend test.

A. K. Pandouria (✉) · P. Chakraborty · V. Tiwari
Indian Institute of Technology Delhi, New Delhi, India
e-mail: purnashis.chakraborty@am.iitd.ac.in; tiwariv@am.iitd.ac.in

S. Kumar
Delhi Technological University, Delhi, India

S. Xia et al. (eds.), *Fracture, Fatigue, Failure and Damage Evolution, Volume 3*, Conference Proceedings of the Society for Experimental Mechanics Series, https://doi.org/10.1007/978-3-030-60959-7_3

3.2 Material and Specimen Preparation

A three-point bend specimen (TPB) is widely used to evaluate fracture toughness of high-strength materials. The material under evaluation for this chapter is Al6063-T6. The chemical composition and static mechanical properties of Al6063-T6 are presented in Tables 3.1 and 3.2.

The specimen for these experiments is designed according to standard ASTM E399 standard [7]. A schematic diagram of the three-point bend specimen (TPB) is shown in Fig. 3.1. The Crack length, a, is nominally between 0.45 and 0.55 times the width, W. TPB specimens can have a width to thickness (W/B) ratio of $1 \leq W/B \leq 4$. The specimen dimensions used in this study are: thickness (B) =10 mm, width (W) = 20 mm, span (S) = 80 mm, and total length (L) = 100 mm. It should be noted that before conducting the experiments, fatigue precrack was done in L-T direction of the specimen, so that the total crack (Notch + Fatigue Crack) should be 50% of the width of the specimen.

3.3 Experimental Procedure

3.3.1 *Quasi-Static Test*

Static three-point bend tests were performed on a zwick/Roell Universal Testing Machine (UTM). The applied load was measured using a load cell of 100kN capacity, and the crack mouth opening displacement was measured using the 3D DIC technique. A detail of the static experimental setup coupled with two low-speed cameras is shown in Fig. 3.2.

The static stress-intensity factor (SIF) for three-point bend specimen is calculated using the following formula [3.1]:

$$K_1 = \frac{3S\sqrt{a}}{2BW^2} Y\left(\frac{a}{W}\right) F \tag{3.1}$$

where

$$Y\left(\frac{a}{W}\right) = \frac{1.99 - \frac{a}{W}\left(1 - \frac{a}{W}\right)\left(2.15 - 3.93\frac{a}{W} + 2.7\left(\frac{a}{W}\right)^2\right)}{\left(1 + 2\frac{a}{W}\right)\left(1 - \frac{a}{W}\right)^{\frac{3}{2}}}$$

Table 3.1 Mechanical properties of Al6063–T6 alloy

Young's modulus E, GPa	Yield strength σ_{ys}, MPa	Ultimate strength σ_{UTS}, MPa	Poisson's ratio	Elongation at break δ, %	Mass density ρ, kg/m^3
68.9	214	241	0.33	12	2700

Table 3.2 Average chemical composition (Wt. %) of Al6063-T6 alloy

Element	Al	Si	Cu	Mg	Fe	Zn	Ni	Mn	Ti
% Wt.	98.55	0.431	0.0069	0.5431	0.3982	0.0153	0.0058	0.014	0.0056

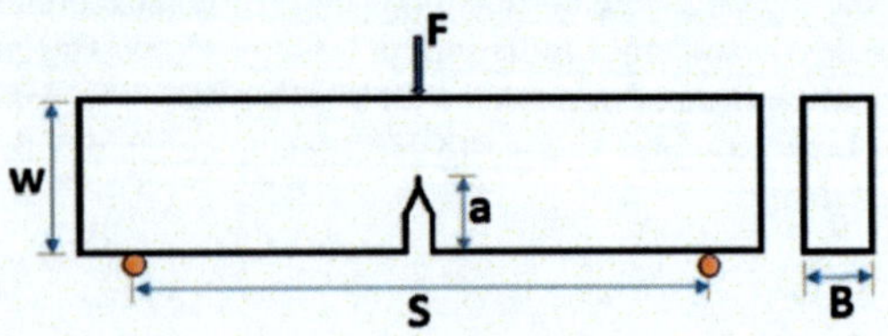

Fig. 3.1 Schematic of 3-point bend specimen

Fig. 3.2 Quasi-static 3-point bend experimental setup

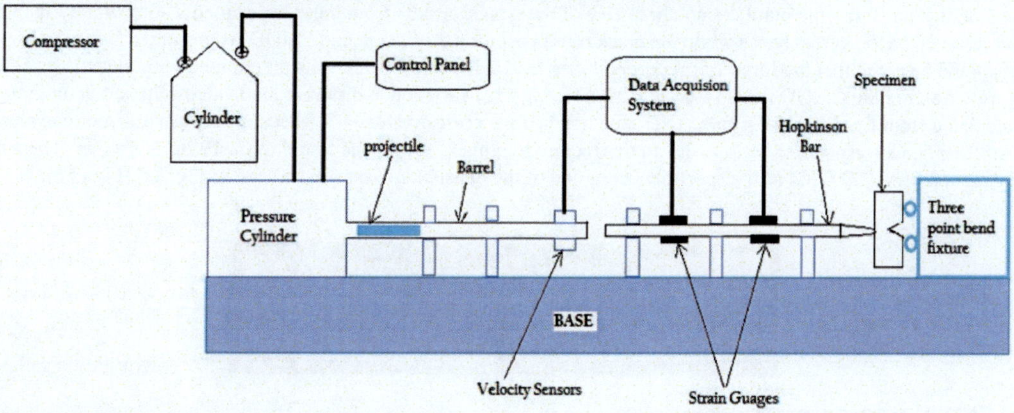

Fig. 3.3 Schematic diagram of MHPB setup used in this study

3.3.2 Dynamic Test

The schematic of the modified Hopkinson pressure bar (MHPB) used in this study [8] is shown in Fig. 3.3. In recent times, it has emerged as most common apparatus to determine the dynamic fracture toughness under high strain loading for three-point bend specimens. It consists of several components such as storage cylinder, barrel, high strength steel bars, projectile, and a 3-point bend fixture. A pulse shaper is placed between the projectile and incident bar to provide a more suitable incident pulse and thereby reducing three dimensional effect. The axial strain was recorded at the two locations on the incident bar using the strain gauges.

3.3.3 Dynamic Stress Intensity Factor (DSIF)

The DSIF can be evaluated from the CMOD by assuming that the relationship of the static case also applies to the dynamic one [3]. In this case dynamic stress intensity factor is expressed by the relation

$$K_1(t) = \frac{E}{4}\frac{w(t)}{\sqrt{a\alpha}}\frac{k_\beta(\alpha)}{v_\beta(\alpha)} \tag{3.2}$$

Where, α is the ratio of crack length and width, β is span length to width ratio, $k_\beta(\alpha)$ is the function that used by Rubio et al. [3] and $v_\beta(\alpha)$ is a non-dimensional function depending on and β values, and its expression can also be found in article published by Guinea et al. [9]. For the case β = 4, $v_\beta(\alpha)$ is given by

$$v_\beta\,(\alpha) = 0.76 - 2.28\alpha + 3.87\alpha^2 - 2.04\alpha^3 + \frac{0.66}{(1-\alpha^2)} \tag{3.3}$$

3.4 Result and Discussion

3.4.1 *Quasi-Static Results*

Quasi-static tests were conducted on a notched TPB specimens of Al6063-T6 with fatigue precrack. One of the deformed specimens can be seen in Fig. 3.4;, it failed within 1.6 mm crack mouth opening. It should be noticed that the load continuously increases up to 1.6 mm crack mouth opening displacement. This increase in load may be due to the large plastic zone that has formed around the notched. One of the typical results from these experiments is shown in Fig. 3.5; here load is presented on the vertical axes while the crack mouth opening displacement (CMOD) is presented along the horizontal axis. It should be noted that the force was recorded using load cell while CMOD was determined with the help of 3D DIC.

Figure 3.5 shows the CMOD (total value of CMOD is U_B-U_A) variation at different load values. Static fracture toughness was calculated from Eq. (3.1) when force (F) value reached the critical value; a 5% secant line was utilized to evaluate this value at 2376 N. Corresponding to this, the static fracture toughness value was found to be 19.62 $MPa\sqrt{m}$. The full field displacement profile (3D DIC) of the specimen from one of the quasi-static test can be seen in Fig. 3.5 (Fig. 3.6).

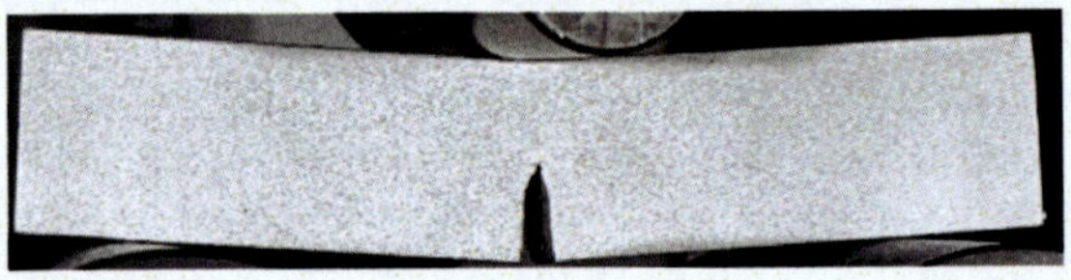

Fig. 3.4 Fractured specimen at low strain rate

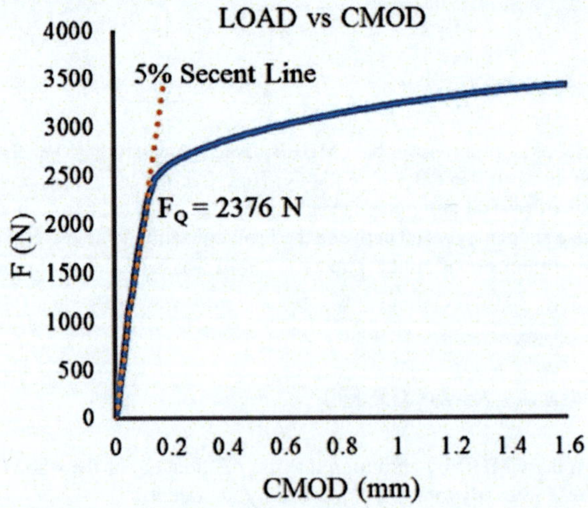

Fig. 3.5 Applied force variation with CMOD

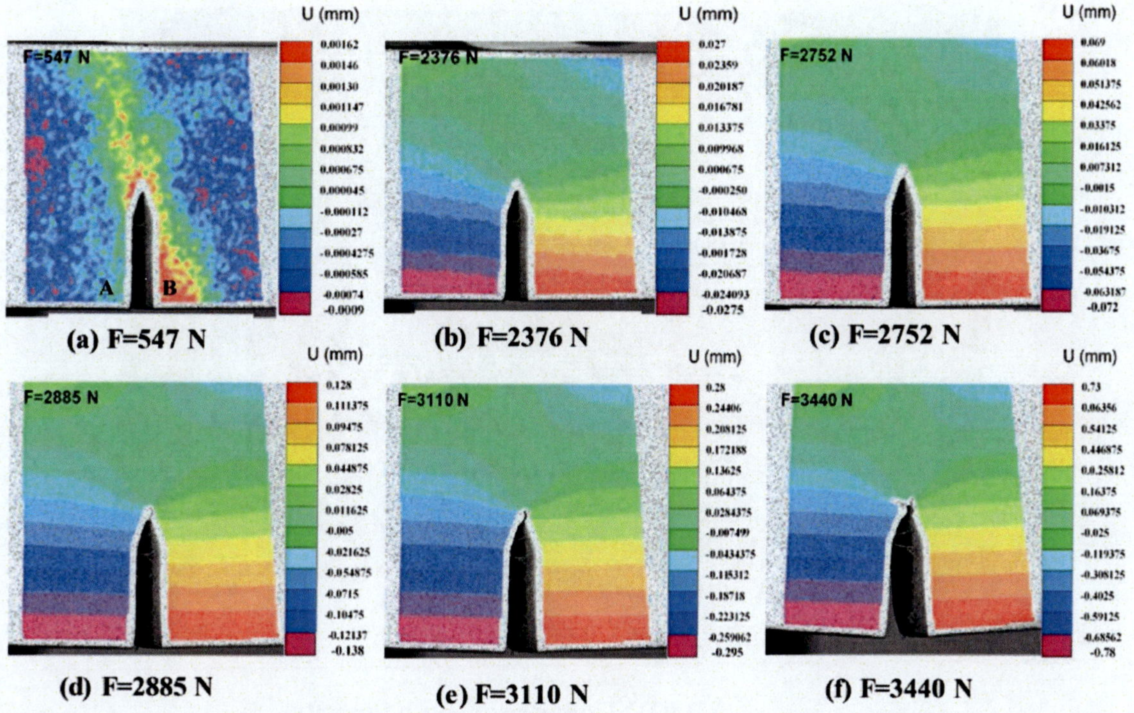

Fig. 3.6 Visualization of CMOD at different static load

3.4.2 Dynamic Experimental Results

The three-point bend experiments were performed under dynamic loading using a MHPB at two different impact velocities (14.1 m/s & 15.95 m/s) in conjunction with 3D digital image correlation (DIC). The specimen was sprayed with white paint, and then a random speckle pattern was marked on it. In the stereo DIC technique, two high-speed cameras (Photron SA5) were placed in front of the specimen facing the speckle pattern. The whole experimental setup synchronized with the high-speed cameras and high-contrast light, as shown in Fig. 3.7. The cameras were programmed to capture images at the frame rate of 100,000 per second and were triggered using the sensors mounted on the incident bar. Camera calibration was performed for each experiment using a 5-mm-spaced calibration grid.

During the experiments, the crack mouth opening displacement (w) was directly measured on the surface of the specimen using 3D DIC, the results of which can be seen in Fig. 3.8. According to this figure, the CMOD value increases continuously during the dynamic loading as well as with the increase in velocity, as expected. The CMOD value is very less because we consider only one wave, which is sufficient to calculate the dynamic fracture toughness at the time of crack initiation. Therefore, data only up to 200 μs are presented for the analysis. Important parameters required for the evaluation of dynamic fracture toughness are stress intensity factor (SIF) and crack initiation time. Dynamic SIF (DSIF) is calculated from Eq. (3.2) by using the crack mouth opening displacement, as shown in Fig. 3.9. Since the DSIF is directly proportional to the CMOD, the behavior of SIF and CMOD curve is almost similar.

Crack initiation time was calculated from the DIC, which is approximately 60 μs for both the cases, as shown in Figs. 3.10 and 3.11. These figures show the full-field transient crack mouth opening displacement w(t) experienced by the specimen during the high strain test at velocities 14.1 m/s and 15.95 m/s, respectively. They also show the deformation of the specimen at different time intervals from which it was observed that the crack initiated at 60 μs. At the crack initiation time, the stress intensity factor becomes the dynamic fracture toughness, which was found to be 21.64 $Mpa\sqrt{m}$ and 25.15 $Mpa\sqrt{m}$ at the considered striker velocities.

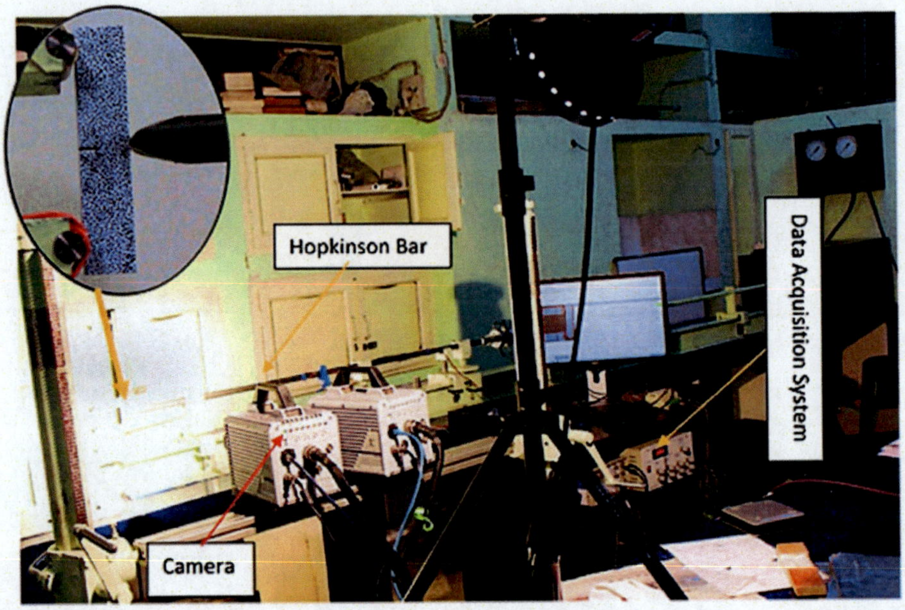

Fig. 3.7 Experimental setup (MHPB) with high speed Cameras

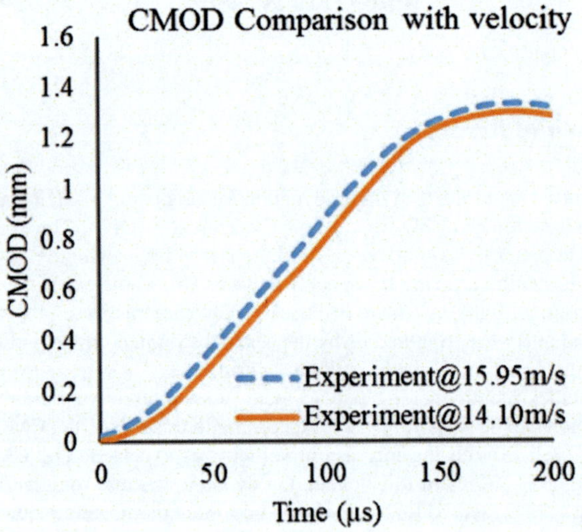

Fig. 3.8 Time response of CMOD with different velocity

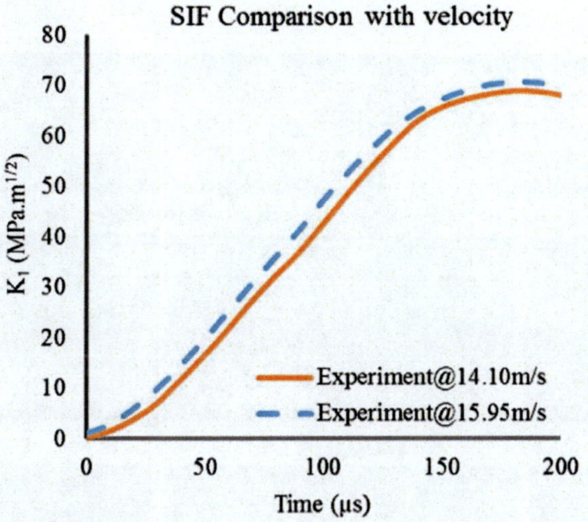

Fig. 3.9 Time response of SIF with different velocity

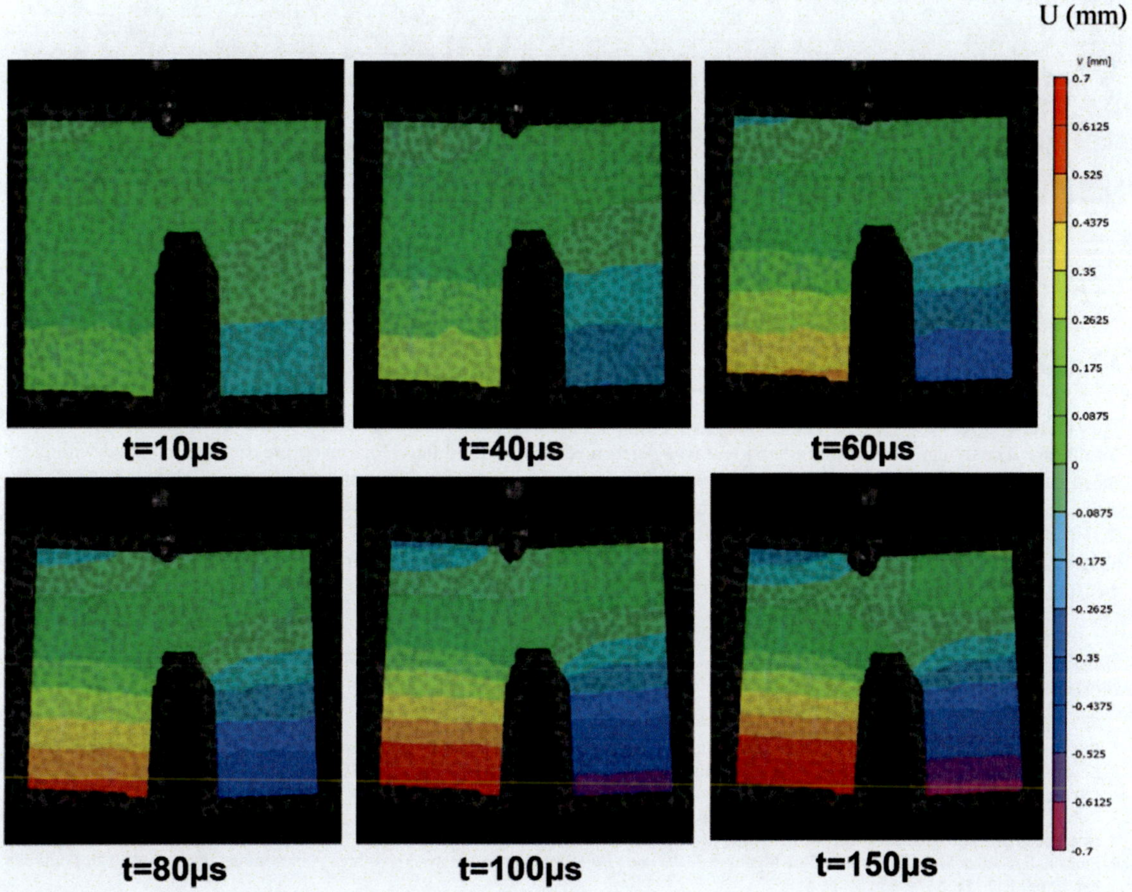

Fig. 3.10 Failure visualization of CMOD at projectile velocity of 14.01m/s

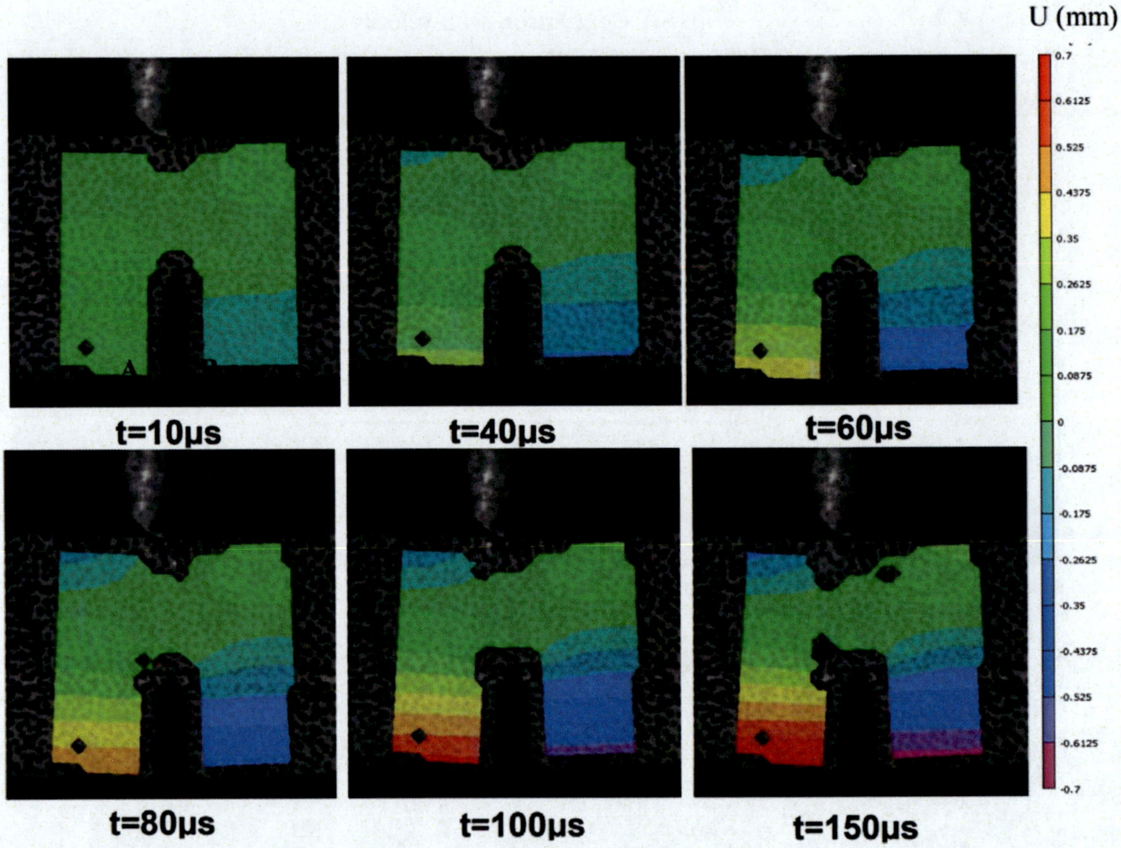

Fig. 3.11 Failure visualization of CMOD at projectile velocity of 15.95m/s

3.5 Conclusion

This chapter presents a failure visualization of Al6063-T6 three-point bend specimen under both static and dynamic conditions. The dynamic three-point bend test was performed using a modified Hopkinson pressure bar coupled with high-speed cameras, while the static bend test was performed according to ASTM 399 standards. The digital image correlations technique was used to calculate the crack mouth opening displacement and crack initiation time. The static fracture toughness ($K_{IC} = 19.62\ MPa\sqrt{m}$) was found to be less than the dynamic fracture toughness ($K_{Id} = 23.39\ MPa\sqrt{m}$). It was also observed that the DIC is a robust technique to analyze the full-field data for any fracture surface, which can be used to measure the exact time and displacement of any cracked surface.

References

1. Bacon, C., Farm, J., Lataillade, J.L.: Dynamic fracture toughness determine from load point displacement. Exp. Mech. **34**, 217–223 (1993)
2. Loya, J.A., Fernandez-Saez, J.: Dynamic fracture initiation toughness determination of Al 7075-T651 aluminum alloy. J. Test. Eval. **35**(1), 1–6 (2007)
3. Rubio, L., Fernandez-Saez, J., Navarro, C.: Determination of dynamic fracture-initiation toughness using three-point bending tests in a modified Hopkinson pressure bar. Exp. Mech. **43**, 379–386 (2003)
4. Pierron, F., Sutton, M.A., Tiwari, V.: Ultra high speed DIC and virtual field method analysis of a three point bending impact test on an aluminium bar. Exp. Mech. **51**, 537–563 (2011)

5. Ming-zhi, X.I.N.G., Yong-gang, W.A.N.G., Zhao-xiu, J.I.A.N.G.: Dynamic fracture behaviors of selected aluminum alloys under three-point bending. Defence Technol. **9**, 193–200 (2013)
6. Sutton, M.A., Orteu, J.J., Schreier, H.: Image Correlation for Shape, Motion and Deformation Measurements: Basic Concepts, Theory and Applications. Springer Science & Business Media, Boston (2009)
7. ASTM Standard E 399-90: Standard test method for plain strain fracture toughness of metallic materials. In: Annual Book of ASTM Standards. ASTM International, West Conshohocken (1994)
8. Engineers and Builders: Standard tools and techniques for dynamic characterization of materials. In: Design book (2015) http://www.engineersandbuilders.com
9. Guinea, G., Pastor, J., Planas, J., Elices, M.: Stress intensity factor compliance and CMOD for a general three-point bend beam. Int. J. Fracture. **89**(3), 103–116 (1998)

Chapter 4
Fatigue Life Prediction of Natural Rubber in Antivibratory Applications

Benoit Ruellan, Jean-Benoit Le Cam, Isabelle Jeanneau, and Frédéric Canévet

Abstract Natural Rubber (NR) is the most commonly used elastomer in the automotive industry, thanks to its outstanding fatigue resistance. Typically, NR exhibits a lifetime reinforcement under non-relaxing loadings (Cadwell SM, Merrill RA, Sloman CM, Yost FL, Ind Eng Chem Anal Ed 12:19–23, 1940). Antivibratory parts are generally made of carbon black–filled NR. Experimental tests are generally carried out in order to check their resistance to fatigue loading, which requires a lot of resources (human presence, time, and money). This is the reason why in the last two decades, predictive approaches were developed to supplement the experimental tests and lighten the development procedure. However, all the physical phenomena, especially the effect of temperature on the fatigue life (B. Ruellan, J.-B. Le Cam, E. Robin, I. Jeanneau and F. Canévet, "Fatigue of natural rubber under different temperatures," *Interational Journal of Fatigue*, vol. 124, pp. 544–557, 2019), are not well taken into account, which explains why the mass of the rubber part and the material formulation remain not optimized. Therefore, it is here proposed to improve the prediction methodology by accounting for the effect of temperature in the fatigue behavior of NR, especially for loadings where a fatigue reinforcement due to strain-induced crystallization is observed. A lifetime prediction model has been developed and results of fatigue life predictions are proposed in the case of variable amplitude loadings. Results are found satisfactory, which enables Contitech AVS France better designing rubber parts by more accurately predicting their fatigue life and perfecting their design.

Keywords Natural rubber · Fatigue life prediction · Automotive industry · Temperature effects · Strain-induced crystallization

4.1 Introduction

Contitech AVS France, the antivibratory division of the group Continental, develops antivibratory parts for the automotive industry. They are mostly made with Natural Rubber (NR) since it presents unrivaled fatigue properties, thanks to its ability to crystallize under stretch [1]. These parts aim at damping the chocks induced by the roadway bumps or the engine vibration. They must answer various specifications provided by car manufacturers. Among them, fatigue ones appear as the most critical in the antivibratory field. Typically, they have to sustain a fatigue loading without failing. Several technical solutions, which implies modification of the design, material recipe, or process parameters, are successively proposed and tested experimentally to improve the fatigue properties, until the specification is met.

However, such an iterative process requires a lot of resources, especially time for producing the prototypes and performing the fatigue tests. This is the reason why in the last two decades, predictive approaches have been developed to supplement the experimental tests and therefore to accelerate and lighten the development procedure. Such methodologies require a good knowledge of the material and a numerical fatigue life prediction model.

The temperature in the engine environment is generally around 90°C; this affects the mechanical behavior of NR by modifying the constitutive behavior and reducing its ability to crystallize [2]. Predictive approaches are generally able to predict the crack initiation location but fail in predicting when it will appear. As a result, the part design is conservative to

B. Ruellan (✉) · I. Jeanneau · F. Canévet
Contitech AVS France, Rennes Cedex, France
e-mail: benoit.ruellan@vc.contitech.fr; isabelle.jeanneau@vc.contitech.fr; frederic.canevet@vc.contitech.fr

J.-B. LeCam
Université de Rennes 1, Institut de Physique UMR 6251 CNRS/Université de Rennes 1, Rennes Cedex, France
e-mail: jean-benoit.lecam@univ-rennes1.fr

S. Xia et al. (eds.), *Fracture, Fatigue, Failure and Damage Evolution, Volume 3*, Conference Proceedings of the Society for Experimental Mechanics Series, https://doi.org/10.1007/978-3-030-60959-7_4

make sure the rubber part sustains the loading. This induces a high mass and cost for the parts, which is not optimal in the industrial context. In this chapter, it is proposed to account for the effect of temperature on the fatigue life of NR to affine the predictive approach. First, a state of the art is presented and justifies the present study. Then, the experimental setup is described before the fatigue behavior is commented for different test temperatures and over a large loading range. A lifetime prediction model has been developed. Unfortunately, it cannot be exposed due to confidential reasons. However, a fatigue correlation is proposed to address the ability of the lifetime prediction model to estimate the number of cycles at end-of-life of Diabolo samples under variable amplitude loadings. Finally, conclusions close the chapter.

4.2 State of the art on the fatigue of NR

NR is among the best candidates for antivibratory applications, thanks to its remarkable fatigue resistance. Typically, the fatigue life of NR is enhanced when the minimum of the loading is kept positive (i.e., under nonrelaxing loadings). This is commonly referred to as the fatigue life reinforcement. This very peculiar phenomenon was observed as early as 1940 by Cadwell and coworkers [1]. Since it has not been observed in noncrystallizable rubbers [3, 4], it is commonly attributed to strain-induced crystallization (SIC).

However, the ability of NR to crystallize under stretch is highly reduced when the temperature is increased [2]. Considering static loadings, Treloar estimated that the crystallites would melt between 75 and 100°C [5]. The significant improvement of crystallinity measurements by Wide Angle X-ray Scattering (WAXS) after the years 2000 enabled more accurate estimations of this temperature: 75°C [6], 80°C [7, 8], 100°C [9], even for large strains. Therefore, it can be considered that the crystallinity is close to zero at 90°C in the antivibratory strain range ($\varepsilon < 200\%$) and for the classical formulations used in automotive applications. Furthermore, it is still not clear what amount of crystallites is sufficient to induce a reinforcement [10, 11]. Note that these results are not necessarily transposable to fatigue loadings [12].

The effect of temperature on the fatigue life of NR has been widely investigated under relaxing loadings. Generally, a decrease in fatigue life occurred when the temperature was increased. This result has been observed on both plane [13] and volumetric samples [1, 14, 15]. However, other experiments on Diabolo samples showed that the fatigue life was only altered from a certain temperature threshold [16]. Only few studies investigated the effect of temperature under different loading ratios, although it could provide information of paramount importance on the role of SIC on the lifetime reinforcement. Bathias et al. showed that a lifetime reinforcement still occurs at 80°C on Diabolo samples but requires more important σ_{mean} than at 23°C [17]. In his PhD thesis, Le Chenadec concluded that the lifetime reinforcement decreased as the temperatures increased from 5 to 100°C. However, only one maximum loading was applied for each loading ratio R [16]. Consequently, the results were extrapolated over the loading amplitude range, rather than being experimentally characterized. Still, these two results seem to indicate that the reinforcement effect decreases with temperature, but it is not clear from which temperature and loading it disappears. This is all the more true as the fatigue resistance may strongly differ from one material formulation to another. Therefore, each material formulation has to be characterized under various temperatures and exhaustive loading conditions.

4.3 Experimental Setup

4.3.1 Material and Sample Geometry

The material here used is a carbon black–filled NR (*cis*-1,4 polyisoprene) vulcanized with sulfur. Diabolo samples are considered for fatigue characterization. Note that more details are provided in [10].

4.3.2 Loading Conditions

The fatigue tests were carried out with a uni-axial MTS Landmark equipped with a homemade apparatus. The latter enables to test independently and simultaneously eight Diabolo samples, which accelerates the fatigue characterization and compensates the inherent dispersion in fatigue lives. The tests were prescribed under displacement, and the local strain was calculated by Finite Element Analysis (FEA) at the sample surface, in the median zone. A Servathin heating chamber was used to investigate the effect of temperature on the fatigue behavior, and three temperatures were applied: 23, 90, and 110°C. Furthermore, a pyrometer recorded the temperature of the material point located in the Diabolo median surface. For that purpose, a second homemade apparatus was developed to provide a suitable kinematics to the pyrometer. The test frequency was chosen in such a way that the global strain rate $\dot{\varepsilon}$ was constant (between 1.8 and 2.4 s^{-1}) from one test to another. In practice, the frequency ranged between 1 to 4 Hz, depending on the strain amplitude. Seven loading ratios ($R = \varepsilon_{min}/\varepsilon_{max}$) were used: −0.25, 0, 0.125, 0.25, 0.35, 0.55, and 0.8. It is here recalled that loading ratios inferior, equal, and superior to zero correspond to tension-compression, repeated tension, and tension-tension, respectively.

4.3.3 End-of-Life Criterion

The crack initiation approach is here used. The end-of-life criterion is based on the maximum reaction force F_{max} evolution. Three regimes were distinguished during the fatigue tests: the softening corresponding to a decrease in F_{max}, the stabilization, and its drop. Note that the drop can be brutal or softer, depending on the loading and the environmental condition. The number of cycles at crack initiation is denoted N_i and refers to the transition between the stabilized regime and the drop in F_{max}. In practice, it has been shown that this coincides with the occurrence of a macroscopic crack with a length not greater than 5 mm at the sample surface.

4.3.4 Scanning Electron Microscopy

Fracture surfaces of failed Diabolo samples were observed with a JSM JEOL 7100F scanning electron microscope (SEM). In addition, the SEM was coupled with an Oxford Instrument X Max Energy Dispersive Spectrometer of X-rays (EDS) and an Aztec software in order to determine the surface fracture composition, especially the precursor nature. The fracture surfaces to be analyzed were beforehand metallized by vapor deposition of an Au-Pd layer.

4.4 Effect of Temperature and Loading on the Fatigue Behavior

As discussed in the introduction, it is proposed to account for the effect of temperature on the fatigue life or NR in order to provide a finer predictive approach in the design loop of antivibratory parts. This section is dedicated to the presentation of the fatigue results under various loadings and temperatures, on an experimental point of view. The Haigh diagram is systematically used to represent the results. Such a diagram gives the influence of the loading on the fatigue life through iso-lifetime curves in the loading amplitude – mean loading space. More details are given in reference [10], especially on the crystallization effects on damage mechanisms. First, the results of reference are provided for tests carried out at 23°C, then the effect of temperature is addressed in a second part, before a discussion is proposed in a third part.

4.4.1 Fatigue Results at 23°C

The iso-lifetime curves are provided with respect to the loading conditions in the Haigh diagram presented in Fig. 4.1 (see the plane lines). The quantity here used to characterize the loading is the force. Discussions on the quantity to use in the context of lifetime prediction are provided in [10]. The black squares represent the experimental fatigue tests according to the stabilized amplitude of force and mean force, in ordinate and abscissa, respectively. They are associated with mean fatigue lives, calculated from eight individual tests. The diagram covers fatigue lives from 10^4 to 10^6 cycles, with $N_1 < N_3$ (graphs with supplementary iso-lifetimes are given in [10]).

At 23°C, the iso-lifetime curves decrease monotonously under relaxing loadings (i.e., -0.25 < R_F < 0). This indicates that the maximum loading applied during a fatigue cycle drives the fatigue life. Under nonrelaxing loadings (i.e., $R_F \geq 0$), the slope of the iso-lifetime curves increases in absolute value and becomes positive for $R_F \geq 0.125$. This means that for a given F_{amp}, increasing F_{mean} improves the fatigue life while F_{max} also increases. This result illustrates the classical fatigue life reinforcement observed under nonrelaxing loadings. Finally, it can be noticed that the reinforcement is less pronounced for $R_F > 0.125$.

The failure surface of Diabolo samples was systematically analyzed at the micro- and macroscopic scale. The analysis revealed the presence of SIC markers: wrenchings, fatigue striations, and cones, which confirmed the activation of SIC under nonrelaxing loadings. More details on damage analysis and SIC markers are provided in [10, 18].

4.4.2 *Effect of Temperature*

Fatigue experiments have also been performed at 90 and 110°C to evaluate the effect of temperature on the fatigue behavior, and especially the lifetime reinforcement under nonrelaxing loadings identified at 23°C. Since the crystallinity decreases with temperatures [2] up to a point that it can be considered as null at 90°C even for large strains, no lifetime reinforcement is here expected under fatigue loadings. In Fig. 4.1, the iso-lifetime curves in dashed and dotted lines correspond to tests performed at 90 and 110°C, respectively.

At 90°C, the curves of the iso-lifetime curves surprisingly change under nonrelaxing loadings (i.e., $R_F > 0$). This indicates that a lifetime reinforcement still occurs at such temperature. Note that the iso-lifetime curve slopes are negative for $R_F > 0.33$, suggesting that the reinforcement vanishes. Such a result had already been reported in the pioneering work by Cadwell and coworkers [1], but for much severe loadings and at 23°C. It could indicate that the detrimental effect of damage induced by the high maximum loadings becomes more important than the beneficial effects of the reinforcement. Finally, no SIC marker was observed on the failure surface of the Diabolo samples, which also goes in the sense that the crystallinity is reduced at 90°C. Since the activation of SIC is evidenced by the presence of the lifetime reinforcement, this result also means that SIC markers require a certain level of crystallinity to form.

At 110°C, the iso-lifetime curves decrease monotonously, regardless of the loading ratio. This suggests that the material behaves as a noncrystallizable rubber and that F_{max} drives the fatigue damage.

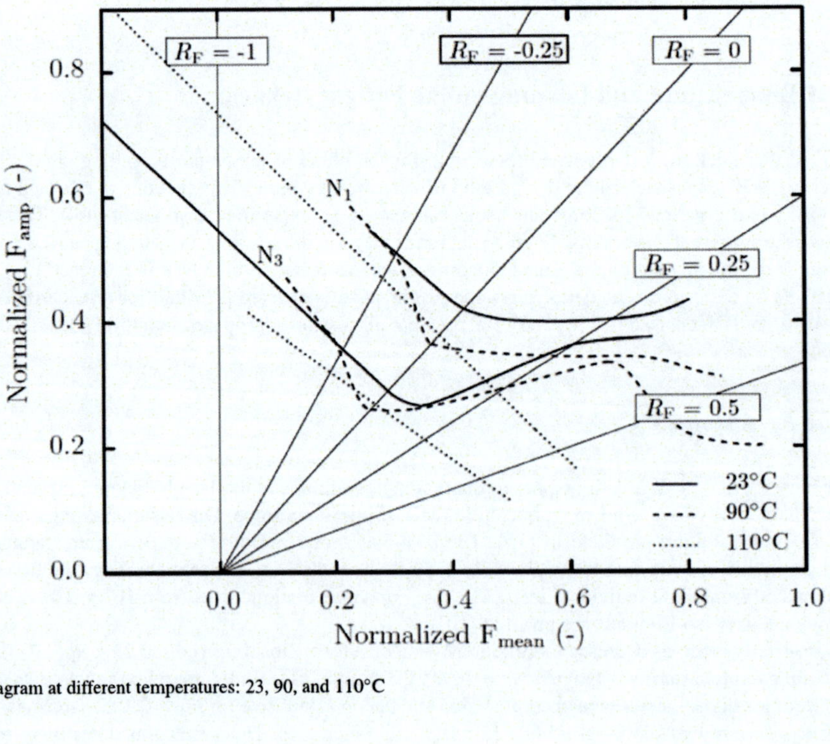

Fig. 4.1 Haigh diagram at different temperatures: 23, 90, and 110°C

4.4.3 Discussion

Even though the lifetime reinforcement is reduced at elevated temperature due to the crystallite melting, it appears that SIC effects still occur. It surely opens discussions on the crystallinity level required to induce such a reinforcement: Is the sole presence of few crystallites able to delay crack initiation? Furthermore, since the crystallinity is generally considered as null at 90°C under quasi-static loadings at strains below 200%, the mechanisms of crystallite growth and melting under fatigue loadings remain to be explored.

Considering fatigue life prediction, since temperature significantly affects the fatigue life under nonrelaxing loadings, it appears crucial to take this effect into account into modeling the fatigue life. As discussed in the introduction, this will provide a finer predictive approach and help antivibratory part developers to answer car manufacturers specification and better designing the parts.

Based on these considerations, a lifetime prediction model denoted "*Rbloc*" has been developed; it accounts for the effect of temperature and loading ratio. It is here kept confidential due to industrial secrecy; however, a fatigue life correlation on Diabolo samples is presented in the next section in the case of variable amplitude loadings.

4.5 Fatigue Life Correlation

4.5.1 Fatigue Life Correlation Conditions

The fatigue life correlation is performed by comparing the experimental fatigue life ($N_{\mathrm{i,exp}}$) with the predicted one ($N_{\mathrm{i,calc}}$). The prediction is carried out by calculating a damage quantity D with the linear Miner's rule as $D = \sum 1/N_{\mathrm{i},j}$, with $N_{\mathrm{i},j}$ the number of cycles at crack initiation for the loading condition j obtained by the experiment. Then, the calculated fatigue life $N_{\mathrm{i,calc}}$ corresponds to the number of cycles for which $D = 1$. The experimental fatigue life is given by the number of cycles corresponding to the drop in stiffness, according to the end-of-life criterion. For each loading condition, the experiment was carried out on 8 Diabolo samples and the mean fatigue life was considered in the analysis.

Two variants of *Rbloc* are here used to demonstrate the need of accounting for the lifetime reinforcement (i.e., the effect of the loading ratio). Variant #1 is the reference. It does not consider the lifetime reinforcement since the fatigue lives $N_{\mathrm{i},j}$ used to access D are calculated based on data obtained at $R = 0$, regardless of the real loading ratio experimentally experienced. In other words, Variant #1 predicts the fatigue life by only using $\varepsilon_{\max}$ as the damage predictor. On the contrary, Variant #2 indexes the fatigue lives based on the experimental R deduced from the input signal. In this case, the prediction combines the relative contributions of R and $\varepsilon_{\max}$.

Three signals defined by blocs of sinus and denoted Signals #1, #2, and #3 were used. Each signal is defined in terms of force by a repetition of 1730 blocs, which corresponds to 121,944 cycles per signal. They are issued from a Rainflow treatment of a road track signal acquired on a torque rod. The force level has been modified to be applicable to Diabolo samples. Typically, the amplitudes of the compression cycles were reduced in absolute value and the force was adjusted to induce a fatigue life in the order of magnitude of 10^5 cycles, to be representative of antivibratory applications. Since *Rbloc* aims at accounting for the lifetime reinforcement under nonrelaxing loadings, the signals are voluntarily highly nonrelaxing.

- Signal #1 is the reference signal (see Fig. 4.2); it covers loading ratios from −0.15 to 0.35, the mean loading ratio being $R_{\varepsilon,\mathrm{mean}} = 0.0865$.
- Signal #2 presents the same maximum loading as Signal #2, but the minimum is increased so that the mean loading ratio increases as well: $R_{\varepsilon,\mathrm{mean}} = 0.1975$. It will be used to address the effect of *Rbloc* to account for the lifetime reinforcement, since the latter is promoted at increasing loading ratio.
- Signal #3 is used to investigate high loading ratios; it covers R from −0.15 to 0.7 with $R_{\varepsilon,\mathrm{mean}} = 0.2131$.

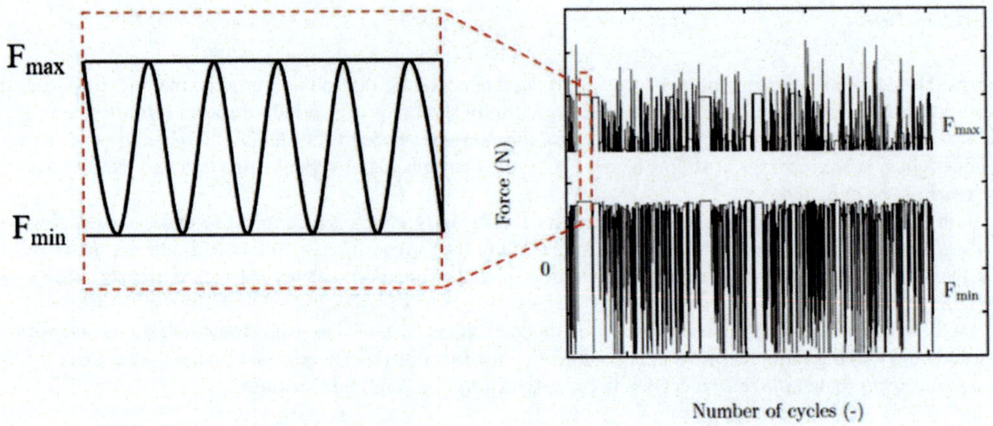

Fig. 4.2 Description of Signal #1, prescribed in terms of force. The zoom-in indicates that the signal is a sinus

Table 4.1 Experimental and calculated fatigue lives for Signals #1, #2, and #3 and for *Rbloc* Variants #1 and #2

Signal		Signal #1	Signal #2	Signal #3
Calculation $N_{i,calc}$	Variant #1 (with $R = 0$)	97,157	97,157	33,940
	Variant #2 (depends on R)	200,332	1,333,081	305,129
Experiment $N_{i,exp}$	Mean	344,325	1,171,700	574,633
	Diabolo 1	410,789	1,221,800	735,769
	Diabolo 2	278,029	–	376,389
	Diabolo 3	–	977,500	679,959
	Diabolo 4	387,309	1,095,900	693,959
	Diabolo 5	373,599	1,217,500	552,659
	Diabolo 6	306,439	1,140,500	–
	Diabolo 7	380,929	1,380,800	370,849
	Diabolo 8	273,179	1,167,900	612,849

4.5.2 Results

The results are given in Table 4.1 and illustrated by the correlation diagram in Fig. 4.3a, b for Variants #1 and #2, respectively. The experimental and calculated fatigue lives are given in ordinate and abscissa, respectively. Individual and mean experimental fatigue lives enable to evaluate prediction accuracy for the three signals.

For Variant #1, the fatigue lives are widely underestimated for the three signals. This could be expected provided the fact that for Variant #1 the fatigue lives are calculated from the $R = 0$ endurance curve, regardless of the loading ratio truly experienced by the Diabolo sample. As a result, it does not account for the lifetime reinforcement under nonrelaxing loadings and the fatigue lives are only obtained from the predictor ε_{max}. This is all the more important that the signals investigated are highly nonrelaxing. Note that Signals #1 and #2 have the same predicted lives since they are defined with identical maximum displacements.

For Variant #2, the fatigue lives are in fairly good agreement with predictions since they are within the factor two scatter band (see the dashed lines in Fig. 4.3). Typically, the benefits of accounting for the lifetime reinforcement can be illustrated by comparing the predicted fatigue lives between Signals #1 and #2. It is recalled that they have similar maximum loadings, but Signal #2 exhibits higher R. It appears quite clear that Variant #2 is able to account for the lifetime reinforcement, which is all the more important that the loading ratio is high. This result clearly highlights the necessity of accounting for the dependence of R on the fatigue life in the fatigue lifetime prediction model.

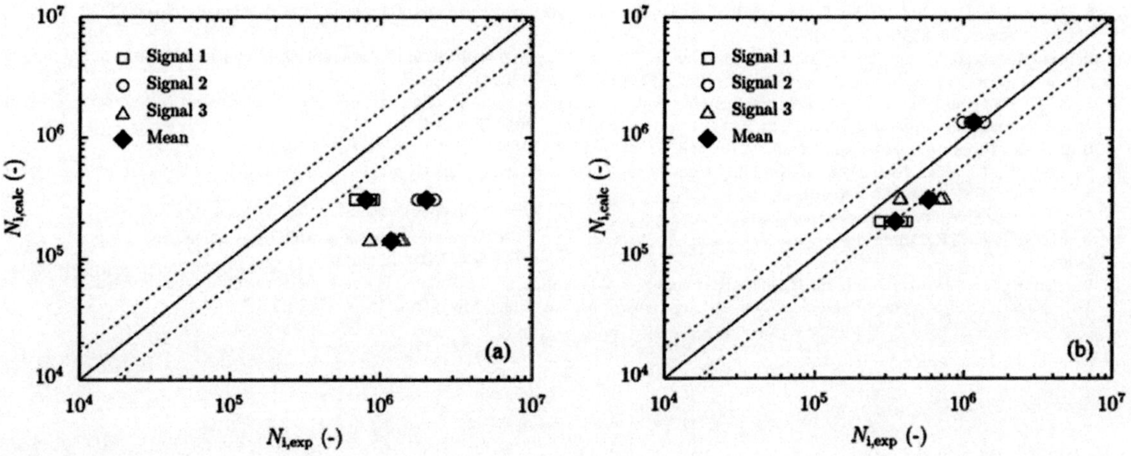

Fig. 4.3 Correlation diagram obtained for the three Signals and for (**a**) Variant #1 (R effect not taken into account) and (**b**) Variant #2 (R effect taken into account)

4.6 Conclusion

Contitech AVS France develops in NR-filled parts for the automotive industry. During the development process of antivibratory parts, predictive approaches are used to supplement classical experimental tests in order to validate their design. A methodology is here proposed to affine and lighten the design procedure by accounting for the effect of temperature on the fatigue behavior. At 23°C, it is here confirmed that the material exhibits a strong lifetime reinforcement under positive loadings ratios. Furthermore, the temperature affects this phenomenon: it is lower at 90°C and disappears at 110°C. Beside opening numerous questions on the effect of temperature on the ability of NR to crystallize, these two results justify the need of accounting for temperature effects into predicting the fatigue life. A lifetime prediction model *Rbloc* has been developed and was here tested on Diabolo samples for blocs of sinus loadings. The predicted fatigue lives were within a factor two scatter band. As a next step, it would be appropriate to validate the lifetime prediction model on industrial parts, which will require to take other physical phenomena into account.

Acknowledgments The authors thank the Contitech France company for supporting this work and for fruitful discussions. The authors also thank the National Center for Scientific Research (MRCT-CNRS and MI-CNRS) and Rennes Metropole for supporting this work financially. SEM images were performed at CMEBA facility (ScanMAT, University of Rennes 1), which received a financial support from the European Union (CPER-FEDER 2007–2014).

References

1. Cadwell, S.M., Merrill, R.A., Sloman, C.M., Yost, F.L.: Dynamic fatigue life of rubber. Ind. Eng.Chem.Anal. Ed. **12**, 19–23 (1940)
2. Trabelsi, S.: Etude statique et dynamique de la cristallisation des élastomères sous tension (2002)
3. Fielding, J.H.: Flex life and crystallisation of synthetic rubber. Ind. Eng. Chem. **35**, 1259–1261 (1943)
4. Lindley, P.B.: Relation between hysteresis and the dynamic crack growth resistance of natural rubber. Int. J. Fract. **9**, 449–462 (1973)
5. Treloar, L.R.G.: The Physics of Rubber Elasticity. Oxford University Press, Oxford (1975)
6. Bruening, K., Schneider, K., Roth, S.V., Heinrich, G.: Kinetics of strain-induced crystallization in natural rubber: a diffusion-controlled rate law. Polymer. **72**, 52–58 (2015)
7. Albouy, P.A., Marchal, J., Rault, J.: Chain orientation in natural rubber, part I: the inverse yielding effect. Eur. Phys. J. E. **17**, 247–259 (2005)
8. Candau, N., Laghmach, R., Chazeau, L., Chenal, J.-M., Gauthier, C., Biben, T., Munch, E.: Temperature dependence of strain-induced crystallization in natural rubber: on the presence of different crystallite populations. Polymer. **60**, 115–124 (2015)
9. Trabelsi, S., Albouy, P.-A., Rault, J.: Stress-induced crystallization around a crack tip in natural rubber. Macromolecules. **35**, 10054–10061 (2002)
10. Ruellan, B., Le Cam, J.-B., Robin, E., Jeanneau, I., Canévet, F.: Fatigue of natural rubber under different temperatures. Int. J. Fatigue. **124**, 544–557 (2019)

11. Rault, J., Marchal, J., Judeinstein, P., Albouy, P.A.: Stress-induced crystallization and reinforcement in filled natural rubbers: 2H NMR study. Macromolecules. **39**, 8356–8368 (2006)
12. Beurrot-Borgarino, S., Huneau, B., Verron, E., Rublon, P.: Strain-induced crystallization of carbon black-filled natural rubber during fatigue measured by in situ synchrotron X-ray diffraction. Int. J. Fatigue. **47**, 1–7 (2013)
13. Duan, X., Shangguan, W.-B., Li, M., Rakheja, S.: Measurement and modelling of the fatigue life of rubber mounts for an automotive powertrain at high temperatures. Proc. Inst. Mech. Eng. Part D J. Automobile Eng. **230**, 942–954 (2016)
14. Beatty, J.R.: Fatigue of rubber. Rubber Chem. Technol. **37**, 1341–1364 (1964)
15. Neuhaus, C., Lion, A., Johlitz, M., Heuler, P., Barkhoff, M., Duisen, F.: Fatigue behaviour of an elastomer under consideration of ageing effects. Int. J. Fatigue. **104**, 72–80 (2017)
16. Le Chenadec, Y.: Autoéchauffement, fatigue thermomécanique des élastomères (2008)
17. Bathias, C., Houel, P., Berete, Y.N., Gorju, K.L.: Damage characterization of elastomeric composites using X-ray attenuation. Fatigue Fracture. (1998)
18. Ruellan, B., Cam, J.-B.L., Robin, E., Jeanneau, I., Canévet, F., Mauvoisin, G., Loison, D.: Fatigue crack growth in natural rubber: the role of SIC investigated through post-mortem analysis of fatigue striations. Eng. Fract. Mech. **201**, 353–365 (2018)

Chapter 5
Fatigue Assessment of Porosity in Electron Beam Melted Ti-6Al-4V

Justin Warner, Dino Celli, Jacob Rindler, M. Herman Shen, Onome Scott-Emuakpor, and Tommy George

Abstract Additive manufacturing (AM) has proven itself to be an effective and versatile solution in replacing aircraft structures and components. However, the AM process still requires the necessary structural reliability as well as the technology to assess operational longevity. In this work, a fatigue performance and damage progression assessment framework is proposed to achieve a fundamental understanding of the fatigue damage mechanisms and its progression in as-built treated electron beam melted (EBM) Ti-6Al-4V at the macroscopic structural scale as well as at the microscopic constituent scale. The work presented utilizes digital image correlation (DIC), an optical strain measurement technique, as a method to detect crack initiation sites occurring on the material's surface and propagating throughout the specimen. A comprehensive testing framework and experimental procedure is developed to generate fatigue data for AM material Ti-6Al-4V as-built specimens. Characterization and simulation of the fatigue progress due to AM process defects (voids, surface roughness, etc.) are also performed using damaging energy progress and damage evaluation.

Keywords Fatigue · Porosity · Computed tomography · Defect damage criterion · Digital image correlation

Many research efforts, within the engineering community, have been focused on the improvement of additive manufactured (AM) materials. Specifically, the fatigue life characterization is of great interest to the aerospace industry due to high standards for the fatigue performance of materials in flight critical components. Powder bed fusion(PBF) is a commonly used AM process to produce components within a small region (200–350 mm) of space by using thermal energy to fuse powder and build a product [1]. There are various options when considering which PBF method to use (i.e., electron beam melting, direct metal laser sintering, selective laser sintering/melting). Electron beam melting (EBM) is an accepted method that is relatively inexpensive and has a small footprint by recycling the unused powder [1]. EBM uses an electron beam source, which can create internal stress-concentrating defects, such as porosity, and significantly reduces fatigue life. This becomes significantly more important for the aerospace industry as fatigue of engineering components made up 25% of all failures and 55% in aircraft components [2]. Therefore, to utilize AM materials for aerospace components, characterizing and ultimately reducing defects within AM materials such as Ti-6Al-4V, Inconel 718, and Aluminum 7075 are demanded to ensure proper safety standards.

Various parameterized defect detriment criterions are investigated to identify the crack initiation site which ultimately leads to fatigue failure. 3D computed tomography (CT) scan data was completed to use identified individual defects characteristics to distinguish fatigue life performance. This was conducted in effort to develop a cycle to failure fatigue life prediction and locates the area of interest of crack initiation prior to destructive testing. Detecting the initiation site can aid in the AM process by informing, via in situ (QM meltpool) or post-CT scans, defect attributes without destructive testing. In addition to determining actual cycles to failure, defect attributes were identified for life prediction with CDM or fracture mechanic fatigue life approaches. Digital image correlation (DIC), an optical strain measurement technique, was used to aid in detection of the location of failure and strains fields. DIC data was utilized to cross-examine the dog bone specimen's regions of interest and evaluate defect detriment criterions' validity. Ti-6Al-4V was analyzed due to its wide application

J. Warner · D. Celli (✉) · J. Rindler · M. H. Shen
The Ohio State University, Columbus, OH, USA
e-mail: warner.799@osu.edu; warner.799@buckeyemail.osu.edu; celli.6@osu.edu; rindler.115@osu.edu; shen.1@osu.edu

O. Scott-Emuakpor · T. George
Air Force Research Laboratory, Wright-Patterson AFB, OH, USA
e-mail: onome.scott-emuakpor.1@us.af.mil; tommy.george@us.af.mil

S. Xia et al. (eds.), *Fracture, Fatigue, Failure and Damage Evolution, Volume 3*, Conference Proceedings of the Society for Experimental Mechanics Series, https://doi.org/10.1007/978-3-030-60959-7_5

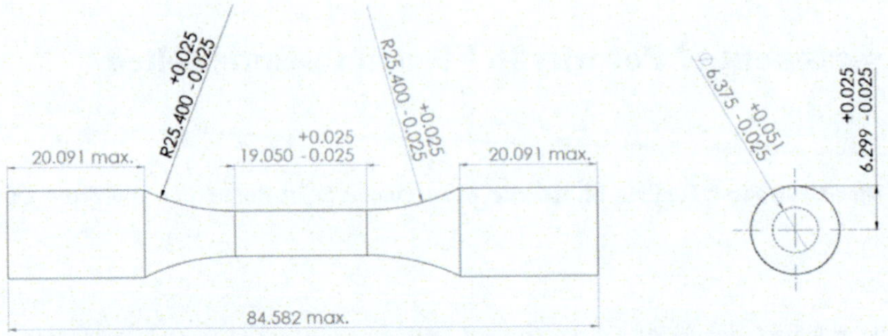

Fig. 5.1 Dimensions of specimen

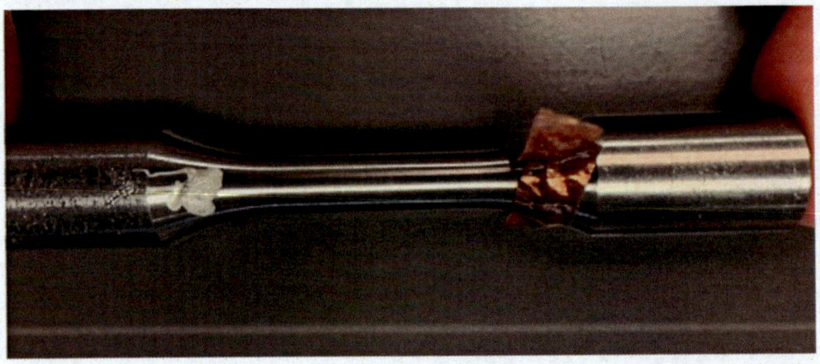

Fig. 5.2 Machined specimen with paste and tape

within the aerospace field. Recent work has shown hot isostatic pressing (HIPing) of Ti-6Al-4V reduces original diameter of void size, thus increasing AM material fatigue life by 20% [3]. Within this document, a porosity analysis has been conducted to examine the effects of the internal voids from a microscopic perspective.

5.1 Method

Materials for testing were obtained from the Center for Design and Manufacturing Excellence (CDME) at The Ohio State University. Ten Ti-6Al-4V cylindrical rods, with a height and diameter of approximately 174 and 13 mm, respectively, were additively manufactured with a vertical build layout using GE Arcam Q10+ Electron Beam system. Built on 1-mm-high supports, the cylindrical rods were randomly distributed throughout the build chamber. The processing parameters were standard OEM Ti-6Al-4V proprietary melt and support themes for control software 5.2.1. The +45-105 μm recycled plasma atomized powder (AP&C) used for the build was all sourced from a singular lot with manufacturer reported chemistry; the number of powder reuses was not tracked (Fig. 5.1).

The rods were machined to the specified parameters shown in Fig. 5.2 with the result shown in Fig. 5.3. CT scans were performed by the Air Force Institute of Technology (AFIT). Specimen labeled 07-TT had a voxel size of 16.49 μm, and all other specimens reported had a voxel size of 14.911 μm. The CT scanner used to obtain the image data sets was the Nikon XT H 225 ST, with the standard settings used. In order to properly orient CT data to DIC, silver conductive paste was applied to the radius of the specimen. Copper tape was applied to the bottom radius for precompensation of beam hardening. CT data was then post-processed in ImageJ, an open architecture software for image data manipulation and reconstruction. The CT slices within the gage section were evaluated and post-processed. Fatigue tests were performed using a 22-kip MTS load frame. Data acquisition for the force data was collected by a MTS FlexTest60 controller. Strain was measured with DIC, and force measurements were taken from the MTS system; this was synchronized to images via a NI USB DAQ model 6212.

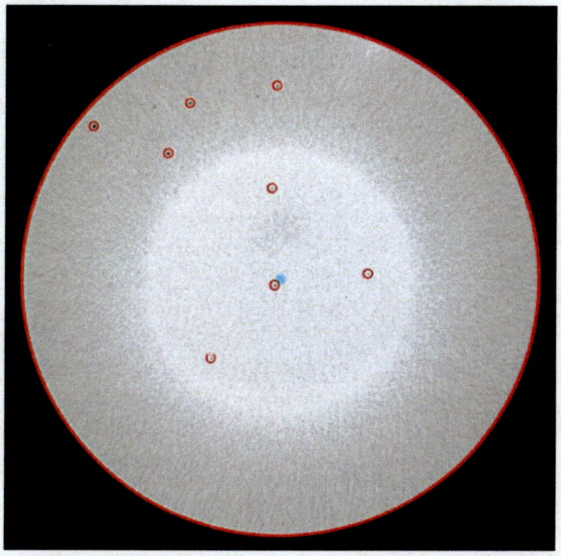

Fig. 5.3 Feature identification

5.2 Results and Discussion

An initial slice was identified for feature identification of the voids and can be seen in Fig. 5.3. The red circles in Fig. 5.3 indicate all the defects found in the image. The blue dot in the center of the image represents the center of the specimen. A border is also placed around the diameter of the specimen which is indicated by a red ring. Although it is best not to alter images to skew digital data, at times, brightness and contrast must be altered to obtain the data needed. However, the Feret diameter can be increased and decreased because of this.

By taking the data from each slice and manipulating the data in MATLAB, the porosity within the entire specimen can be obtained. Figure 5.4 shows the result of a mapping of the porosity radially normal to the z-axis. Figure 5.4 is porosity relative to edge distance in a polar coordinate system from the z-plane of a single specimen. Data of the average porosity size found from ImageJ in Table 5.1 for all as-built specimens.

Tammas-Williams proposed multiple rankings for stress intensity factors; a ranking of interest was selected from the multiple rankings. The failure location was estimated by using a simplistic version of the equation which includes the nonuniform stress due to the porosity position($\sigma_x(position)$), the stress concentration in relation to the surface distance ($K_t(surface)$), the pore aspect ratio ($K_t(AR)$), proximity to other pores ($K_t(proximity)$), and the area of a defect normal to the applied stress ($\sqrt[4]{A_n}$) [4]. The full equation can be seen in Equation 5.1.

$$\Delta K_I(relative) = \sigma_x(position)K_t(surface)K_t(AR)K_t(proximity)\sqrt[4]{A_n} \tag{5.1}$$

It is important to note that all the defects are assumed as round and spherical by shape; none are oblate. Because of this assumption, it can be shown by Biswal the aspect ratio will be 1.0 [5]. Therefore, the reduced equation is written as Equation 5.2 in the following.

$$\Delta K_I(relative) = K_t(surface)K_t(proximity)\sqrt[4]{A_n} \tag{5.2}$$

There are two other methods proposed and evaluated in this study. Equation 5.3 uses portions of the previous values from Equation 5.1 but additionally includes the length of the pore in the z-direction ($K_t(length)$) and the radius (r) of the pore. Therefore, Equation 5.3 considers the extra dimension within the defect damage criterion.

$$\Delta K_I(relative) = K_t(surface)K_t(proximity)K_t(length)r \tag{5.3}$$

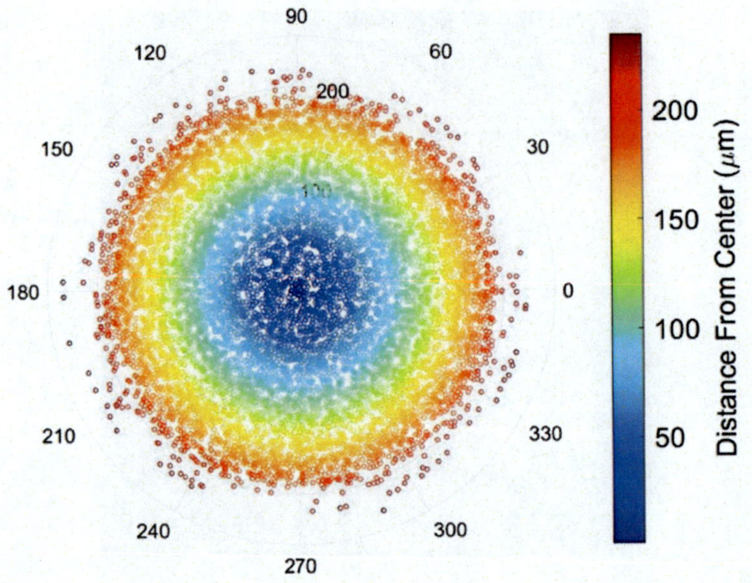

Fig. 5.4 Mapping of porosity

Table 5.1 Specimen porosity and life characteristics

Specimen name	Applied stress (MPa)	Fatigue cycle life (cycles)	Avg. Feret diameter (μm)	Estimated CT slice failure Loc.	CT slice failure Loc. (distance from closest estimation)
01-TT	689	5.08×10^3	58.07	$902^4, 1091^4, 1758^2, 1758^3$	784 (1759.5 μm)
04-TT	449.7	7.82×10^4	54.93	$1062^2, 1062^3, 1899^4, 1916^4$	1616 (4473.3 μm)
07-TB	689	13.02×10^3	57.93	$339^4, 1798^4, 1826^2, 1827^1$	1732 (1088.3 μm)
10-TT	499.5	2.66×10^4	57.95	$328^3, 1002^4, 1166^4, 1597^2$	1764 (2490.1 μm)

Lastly, a facile formula is used to produce a prediction of crack initiation location as Equation 5.4. Equation 5.4 only utilizes the distance from edge (ED) and radius of pore size (r). It is commonly known that defects in relation to the edge directly relate to the detriment of fatigue life. Additionally, defect/porosity size can also be a significant factor at times when determining the location of failure. Therefore, ED and the r will be used for the final prediction criterion.

$$\Delta K_I(relative) = (ED)r \tag{5.4}$$

The results from the fatigue tests can be found in Table 5.1 as well as the average Feret diameter of each specimen, the estimated CT slice locations from each defect damage criterion, and the identified CT slice failure location. The CT slice failure locations in Table 5.1 were found by identifying a strain field located on the specimen within VIC 3D. CT images were evaluated visually and were compared to the fracture surface captured on the microscope. The CT slice which aligned with the location of failure was documented. In Table 5.1 the different criterions are indicated by the estimated CT slice failure location value having a superscript number associated with each equation.

A correction was applied to locating the true CT image where failure occurred from DIC. The original CT slice estimation from DIC for each as-built specimen is shown in the following with the corrected value enclosed in parentheses: 01-TT CT = 797 (784), 04-TT CT = 1619 (1616), 07-TB CT = 1732 (1732), and 10-TT CT = 1750 (1764). The change of CT slice failure location was made by visually inspecting the fracture surface to that of the location CT scan images.

Figure 5.5 shows the fracture surface associated with specimen 04-TT. At the top of the image, a defect can be seen by its bright intensity compared to the rest of the specimen. This defect is the location of the crack initiation and is viewed as a gas pore caused during the EBM process. Shown in Fig. 5.6, the CT scan image of the determined failure location shows the same pore near the top-right edge of the specimen cross-section. The pore at the edge is the same crack initiation pore found from Fig. 5.5. The pore located was also witnessed in DIC and the crack can be seen in Fig. 5.8. The SN curve for the data

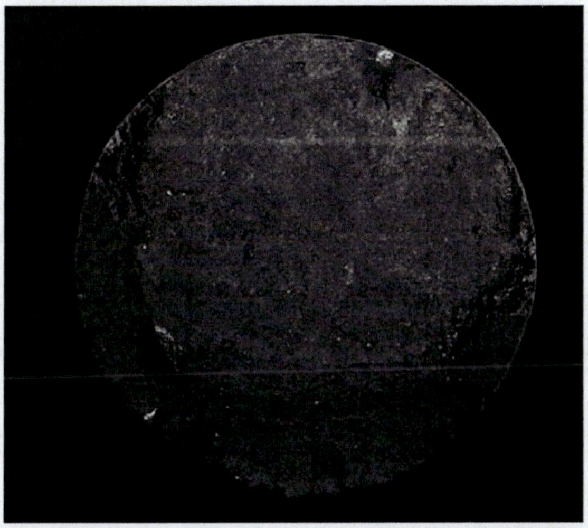

Fig. 5.5 04-TT Fracture surface

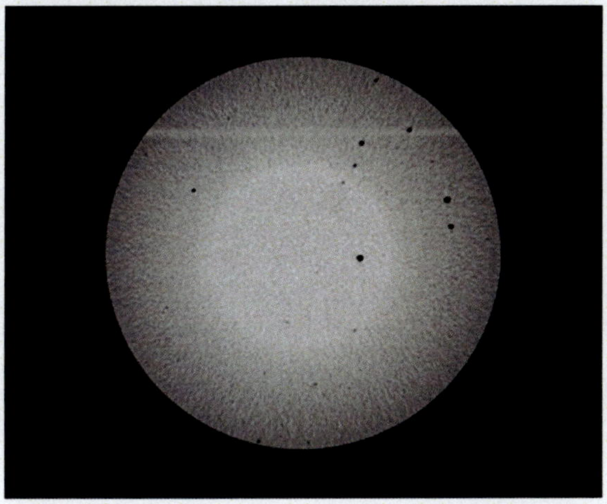

Fig. 5.6 04-TT Fracture CT image

tested is shown in Fig. 5.7. Majority of the specimens tested followed the 10% significance confidence interval presented in compared literature [6].

Shown in Fig. 5.8, specimen 04-TT is presented with a contour of the strain field, determined via DIC, after a crack has initiated. The spectrum of colors indicate the strain on the specimen where dark purple indicates 0.00245 and the dark red indicates 0.0088. However, the scale bar is not presented within Fig. 5.8 because of the discontinuity of the strain field due to the presence of the crack and therefore can only be utilized as a qualitative means of identifying localized deformation. This, however, demonstrates how the location for a crack was found.

CT images of Ti-6Al-4V were obtained and referenced to the speckle images from DIC and the fracture surface of the specimen, and thus a conclusion was made of where the specimen broke as well as the primary defect which ultimately led to crack initiation. From Table 5.1, it was found that Equation 5.4 yielded the closest predicted location of failure specimen 01-TT, 04-TT, and 07-TB and Equation 5.2 yielded the closest prediction for specimen 10-TT. All the prediction methods were able to approximate majority of the specimens failure location by using defect characteristics found from CT images within approximately 1–2 mm.

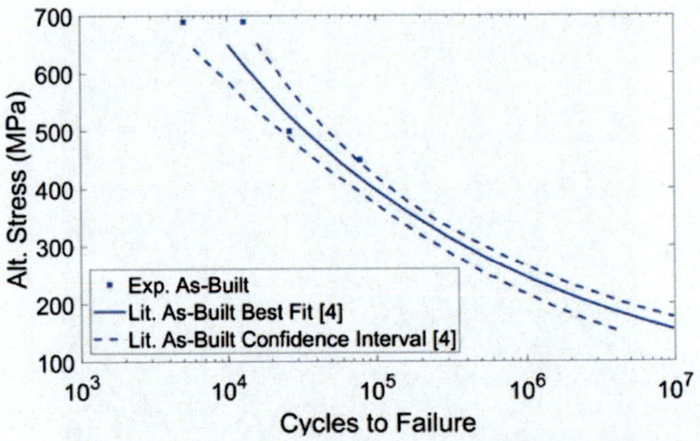

Fig. 5.7 Fatigue life results [4]

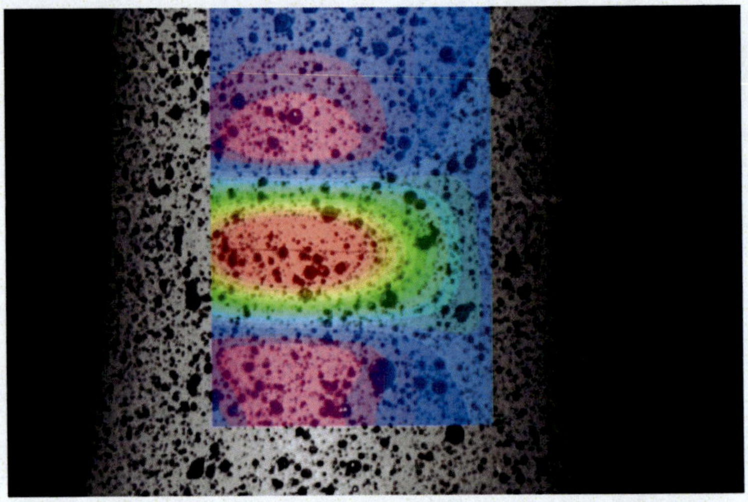

Fig. 5.8 04-TT DIC strain field

5.3 Conclusion

For the limited sample of specimens tested, the fatigue life has yielded similar SN behavior found in the literature and suspected to be of nominal material properties to other EBM as-built specimens. LOF caused a crack initiation for one specimen, causing fatigue life to be lower than expected.

- The location of the crack initiation can be estimated to a specific area, but there are various factors affecting the start of a crack which may not have been considered for future testing.
- By using multiple criterions, a prediction of failure can be found within 5,000 µm at most and is usually around 2,000 µm.
- The algorithms have the capability of improving, but they show great potential for detecting stress fields and failure points.
- LOF was detected throughout the specimens and ultimately caused a crack initiation for one specimen causing fatigue life to be lower than expected. This requires future work to identify LOF within CT images or by a different means to fully analyze all defects within in AM fabricated material.
- Equation 5.3 of the methods used to predict failure location performed the worst out of the three. This ultimately shows that the length does not have a significant impact on the failure location with respect to axial loading.

Acknowledgments The authors would like to thank the Turbine Engine Fatigue Facility of the Air Force Research Laboratory at Wright-Patterson Air Force Base for equipment use, assistance, and technical support of this research effort. The authors would also like to acknowledge and thank the Center for Design and Manufacturing Excellence at the Ohio State University for supplying and creating the material needed with their equipment.

References

1. Tofail, S.A., Koumoulos, E.P., Bandyopadhyay, A., Bose, L., O'Donoghue, S., Charitidis, C.: Additive manufacturing: scientific and technological challenges, market uptake and opportunities. Mater. Today **21**(1), 22–37 (2018)
2. Findlay, S., Harrison, N.: Why aircraft fail. Mater. Today **5**(11), 18–25 (2002)
3. Hrabe, N., Gnäupel-Herold, T., Quinn, T.: Fatigue properties of a titanium alloy (ti-6al-4v) fabricated via electron beam melting (ebm): effects of internal defects and residual stress. Int. J. Fatigue **94**, 202–210 (2017)
4. Tammas Williams, S., Withers, P., Todd, I., Prangnell, P.: The influence of porosity on fatigue crack initiation in additively manufactured titanium components. Sci. Rep. **7**(1), 7308 (2017)
5. Biswal, R., Syed, A.K., Zhang, X.: Assessment of the effect of isolated porosity defects on the fatigue performance of additive manufactured titanium alloy. Addit. Manuf. **23**, 433–442 (2018)
6. Chern, A.H., Nandwana, P., Yuan, T., Kirka, M.M., Dehoff, R.R., Liaw, P.K., Duty, C.E.: A review on the fatigue behavior of ti-6al-4v fabricated by electron beam melting additive manufacturing. Int. J. Fatigue **119**, 173–184 (2019)

Chapter 6
Bayesian Updating of a Cracking Model for Reinforced Concrete Structures Subjected to Static and Cyclic Loadings

Henriette M. Imounga, Emilio Bastidas-Arteaga, Moutou Pitti Rostand, and Serge Ekomy Ango

Abstract Several reinforced concrete structures fail by fatigue loads. The effects of this type of loading are complex. Many mechanical models based on the damage theory could be used to represent the behavior of reinforced concrete under cyclic loading. Their use requires, among others, knowledge of the material characteristic parameters and its related uncertainties that could be determined from experimental tests. However, the models are time-consuming and the experimental data scarce. In this chapter, we propose a Bayesian network–based methodology to propagate uncertainties in the damage theory model. The proposed methodology is useful to identify the uncertainties of the damage model used when some parameters are measured. The methodology is illustrated with a reinforced concrete beam subjected to cyclic loading. The results obtained were compared with those of the experimental tests to validate the proposed methodology. The good agreement indicates that our approach is capable of propagating uncertainties and integrating data from experimental tests. The proposed approach could be also used to identify the uncertainties of the model used by introducing experimental measurements.

Keywords Cracking · Cyclic loading · Bayesian network · Reinforced concrete · Damage theory

6.1 Introduction

During their operation, reinforced concrete structures are subjected to stresses of several types responsible for their rapid deterioration. To reproduce or predict the degradation processes of these structures, numerical models validated by experimental tests have been developed for many authors. In the case of fatigue loads, several models based on the damage theory are useful to describe the mechanical behavior of concrete after each loading cycle (loading-unloading) [1–3]. The purpose of these models is to assess the effects of micro-cracking in concrete, using a bounded variable (scalar or tensor) called the damage variable. For the models of damage, the behavior law relating stress to deformation is written as follows:

$$\boldsymbol{\sigma} = (\boldsymbol{I} - \boldsymbol{D})\,\boldsymbol{C} : \boldsymbol{\varepsilon} \tag{6.1}$$

where σ is the stress tensor, ε the strain tensor, I the unity tensor, C is the tensor of elastic behavior, and D is the tensor of damage. These models take into account intrinsic parameters of the material, such as the elastic modulus, the stresses

H. M. Imounga
Université des Sciences et Technique de Masuku, Franceville, Gabon

UBL, Université de Nantes, Nantes, France
e-mail: henriette.imounga@etu.univ-nantes.fr

E. Bastidas-Arteaga
UBL, Université de Nantes, Nantes, France
e-mail: henriette.imounga@etu.univ-nantes.fr; emilio.bastidas@univ-nantes.fr

M. P. Rostand (✉)
Université de Clermont Auvergne, CNRS, SIGMA Clermont, Institut Pascal, Clermont-Ferrand, France

IRT, CENAREST, Libreville, Gabon
e-mail: rostand.moutou_pitti@uca.fr

S. Ekomy Ango
IRT, CENAREST, Libreville, Gabon
e-mail: ekomyango@yahoo.fr

S. Xia et al. (eds.), *Fracture, Fatigue, Failure and Damage Evolution, Volume 3*, Conference Proceedings of the Society for Experimental Mechanics Series, https://doi.org/10.1007/978-3-030-60959-7_6

of tension/compression, or even the parameters of hardening, which are determined from experimental tests. These tests may contain uncertainties related to the implementation of the tests, the variability of the material properties, the errors in the measurements, and the used calculation methods. In this context, the main objectives of this work is to propose a methodology based on Bayesian networks that could be useful to propagate uncertainties in the damage model and that could be updated with some measurements. The first part of this work describes the constitutive equations of the used damage model (Sect. 6.2). The numerical implementation and the description of the proposed Bayesian network–based methodology are detailed in Sect. 6.3. Finally, the results are presented and discussed in Sect. 6.4.

6.2 Constitutive Equations of the Damage Model

The model used is an isotropic damage model developed by Richard et al. [3] and implemented in Castem codes. In this model, the damage variable representing the degradation of the elasticity modulus, due to micro-cracking, is a scalar which varies between 0 for healthy concrete and 1 for cracked concrete. The constitutive equations are formulated in the framework of the thermodynamics of irreversible processes and the model accounts for residual deformations, hysteretic behaviors, and unilateral effects (opening - closing of the crack) of concrete [3]. The different state laws based on the inequality of Clausius–Duhem–Trusdell [4] are summarized by the following equations:

$$\begin{cases} \boldsymbol{\sigma} = \mathrm{E}((1-\mathrm{D}) < \varepsilon_x >_+ + < \varepsilon_x >_-) + 2(1-D)\,\mu\boldsymbol{\varepsilon} + 2D\mu\,(\boldsymbol{\varepsilon} - \boldsymbol{\varepsilon}^{\pi}) \\ \boldsymbol{\sigma}^{\pi} = 2D\mu\,(\boldsymbol{\varepsilon} - \boldsymbol{\varepsilon}^{\pi}) \\ Y = \frac{E}{2} < \varepsilon_x >_+^2 + \mu\boldsymbol{\varepsilon}.\boldsymbol{\varepsilon} - \mu\,(\boldsymbol{\varepsilon} - \boldsymbol{\varepsilon}^{\pi}).(\boldsymbol{\varepsilon} - \boldsymbol{\varepsilon}^{\pi}) \end{cases} \tag{6.2}$$

where μ is a shear coefficient, $\boldsymbol{\sigma}^{\pi}$ is the frictional tensor, $\boldsymbol{\varepsilon}^{\pi}$ is the sliding tensor, and Y is the damage energy released rate. Equation (6.3) gives the expression of the scalar damage variable:

$$D = 1 - \frac{1}{1 + \left(A_{\mathrm{Dir}} H\left(< \varepsilon_{ij} >_+ < \sigma_{ij} >_+\right) + A_{\mathrm{Ind}}\left(1 - H\left(< \varepsilon_{ij} >_+ < \sigma_{ij} >_+\right)\right)\right)(Y - Y_0)} \tag{6.3}$$

where A_{Dir} represents the brittleness in tension, A_{Ind} the brittleness in compression, Y_0 the damage initial threshold, and H is Heaviside.

6.3 Numerical Implementation and Parameters Updating

6.3.1 Problem Description

The approach is illustrated based on the configuration of a test conducted by Wang et al. [5]. The specimens are reinforced concrete beams of dimensions 300 mm × 120 mm × 1500 mm, subjected to a cyclic bending at three points (10,000 cycles), with an alternating point load varying between 5.4 kN and 18 kN. The geometry of the beam as well as the configuration of the loading is presented in Fig. 6.1.

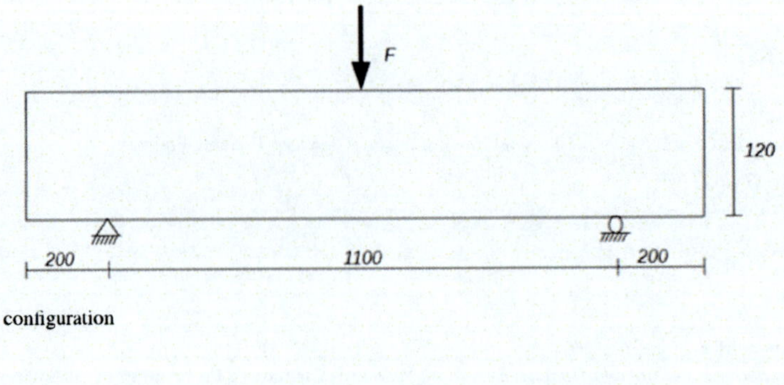

Fig. 6.1 Beam geometry configuration

Table 6.1 Input parameters of mechanical model

Parameters	Notation	Value
Elasticity modulus (MPa)	E	Lognormal (μ = 35,000; COV = 0.12)
Tensile strength (MPa)	f_t	Lognormal (μ = 3.5; COV = 0.13)
Concentrate load (N)	F	Uniform on the interval [5.4; 18] × 10^3
Damage initial threshold (Jm^{-3})	Y_0	1000
Poisson ratio	ν	0.2
Brittleness in tension	A_{Dir}	1×10^{-2}
Brittleness in compression	A_{Ind}	5×10^{-4}
Kinematics hardening (Pa)	γ_0	7×10^{9}
Nonlinear hardening (Pa)	a_0	7×10^{-7}

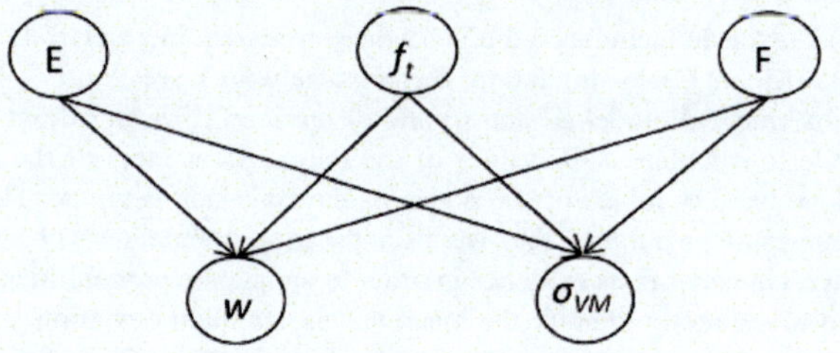

Fig. 6.2 Bayesian network setup

Table 6.2 Discretization of nodes and a priori information for the Bayesian network

Nodes	Number of states	Prior distribution	Boundaries
E (MPa)	20	LN (μ = 35,000; COV = 0.12)	[20; 50] × 10^3
f_t (MPa)	20	LN (μ = 3.5; COV = 0.13)	[1; 5]
F (N)	20	Uniform	[5.4; 18] × 10^3
w (mm)	40	–	[3; 0]
σ_{VM} (MPa)	20	–	[10; 80] × 10^3

6.3.2 Mechanical Model

The beam was modeled in Castem finite element software in 2D case with the assumption of plane stresses. The input parameters of the model are presented in Table 6.1. The Young's modulus, the tensile strength, and the applied load are modeled as random variables whose laws and parameters are also defined in Table 6.1. The other parameters (brittleness in tension and in compression, kinematics, and nonlinear hardening) of the model are taken from [1, 3, 6] for ordinary concretes because we don't have information on this. These parameters are not easy to measure during experimental tests unlike others, such as the deflection whose observations will be used as described in the following sections. One thousand Monte Carlo simulations were performed for each random variable. The outputs of the finite element code are the values of the mid-span deflection (w) and the stress of Von Mises (σ_{VM}) in the direction of loading.

In order to integrate the test data into the Bayesian network (Sect. 6.3.3), the properties (material characteristic, geometry, loading, etc.) of the numerical model are the same as those of the real beam.

6.3.3 Bayesian Configuration and Updating

The configuration of the Bayesian network is presented in Fig. 6.2. It is constructed from the input random variables and output data estimated from the finite element code. The input random variables (E, F, and f_t) are modeled as parent nodes, and the output parameters (w, σ_{VM}) as child nodes. All nodes are considered discrete and divided into a number of states within predefined bounds. The details of the discretization and the a priori information of the different nodes are summarized in Table 6.2.

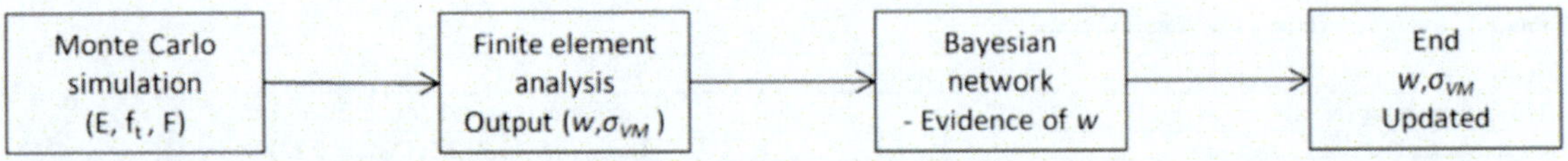

Fig. 6.3 Flowchart of the proposed methodology

Table 6.3 Deflection and tensile stress means

Parameters	No evidence	With evidence	Experimental data
w (mm)	0.53	0.82	0.6
σ_{VM} (MPa)	51.12	61.62	–

The methodology used to update the deflection and the Von Mises stress requires an iteration of mechanical (finite element code) and probabilistic analyses (Monte Carlo simulation, Bayesian network), see Fig. 6.3. Monte Carlo simulations of the input random variables of the mechanical model (E and f_t) and of the load (F), carried out in the MATLAB software, are introduced into the Castem code to calculate 1000 values of the output parameters of the model (w, σ_{VM}). These values allow building the conditional probability tables of the nodes of the Bayesian network. This step allows propagating the uncertainties of the input parameters to the outputs. The data from the experimental tests observed on the mid-span deflection are then introduced into the Bayesian network as evidence in order to update the probabilities of all the nodes. The deflection measurements can also be used to indirectly identify the mean values, standard deviation, and type of distribution of other mechanical parameters. The Bayesian network outputs are the a posteriori probabilities of each of the nodes, from which the laws of the nodes and their parameters (mean, standard deviation) could be determined after updating.

6.4 Results and Discussions

The values of the maximum mid-span deflection before and after the update with the Bayesian network are presented in Table 6.3. The mean deflection value for 10,000 cycles to the unload is also presented in Table 6.3 [5]. We can note that the mean values of w obtained from the Bayesian network are higher than the experimental value; and that obtained without the addition of evidence is closer to experimental value. In addition, the value of the maximum tension stress increases with the introduction of the evidence of the deflection. This difference shows on the one hand that the uncertainties linked to the parameters of the concrete propagate in the mechanical model before the update and on the other hand that they are linked to the experimental evidence propagated through the Bayesian network. Uncertainties on experimental measurements are more important than the uncertainties linked to the parameters.

These preliminary results can be improved by adjusting the limit values, the number of states, or even the number of Monte Carlo simulations, which was fixed at 1000 simulations because of the computation time.

Figure 6.4 shows the deflection histograms with and without evidence. We can see that the a posteriori histogram of w adjusts with the introduction of evidence from the trials. This adjustment is due to the fact to had integrated additional information on the deflection. The proposed methodology therefore makes it possible to update the output parameters of the mechanical model by integrating experimental measurements.

6.5 Conclusion

This chapter proposes a methodology which combines a mechanical model and probabilistic approaches. The methodology makes it possible to account for the uncertainties linked to the material, loading, and experimental measurements. It is also useful to integrate the observations of the tests and to update the other parent nodes by using Bayesian networks. The results show that the uncertainties related to the experimental measurements are greater than those linked to the characteristics of the material in view of the difference between the numerical values of the deflection before and after the introduction of the evidence. Although the number of simulations is small, the numerical values of the deflection are not very far from those of the experimental test. Future work will focus on the analysis of cracking under cyclic load, taking into account the phenomena of opening and closing of the crack. Another aspect of the improvement of the methodology will be the

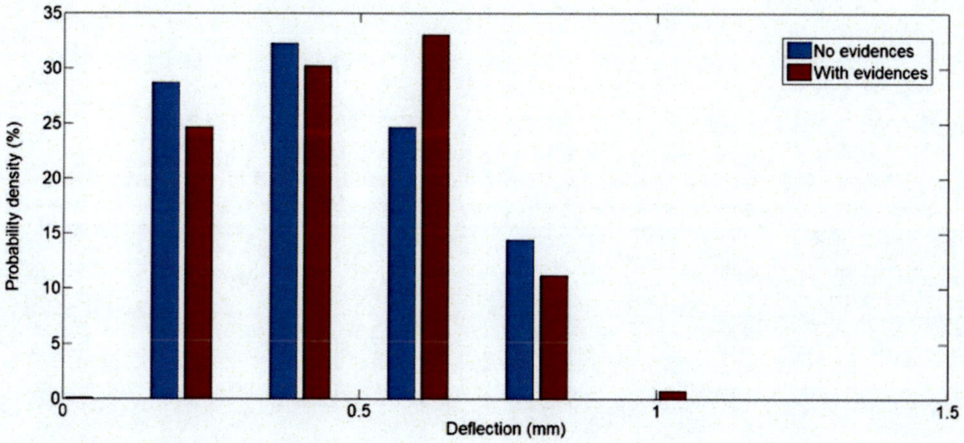

Fig. 6.4 Comparison of a posteriori histograms of w

consideration of the geometry of the crack, the space between the cracks, and the propagation of the fatigue cracks. Finally, the parameters of the damage model whose values are taken from the literature will be modeled as random variables in order to evaluate the effects of their uncertainties on the output parameters of the mechanical model.

Acknowledgments The financial support of the Regional Council of "Pays de la Loire" within the framework of the BUENO 2018–2021 research program (Durable Concrete for Offshore Wind Turbines) is gratefully acknowledged.

References

1. La Borderie, C.: Unilateral effects for damage-like materials: modelling and application to concrete structures. PhD Thesis, University Paris VI, Paris (1991)
2. Matallah, M., La Borderie, C.: Inelasticity–damage-based model for numerical modeling of concrete cracking. Eng. Fract. Mech. **76**(8), 1087–1108 (2009)
3. Richard, B., Ragueneau, F., Cremona, C., Adelaide, L.: Isotropic continuum damage mechanics for concrete under cyclic loading: stiffness recovery, inelastic strains and frictional sliding. Eng. Fract. Mech. **77**(8), 1203–1223 (2010)
4. Lemaître, J., Chaboche, J.L.: Solid Material Mechanics. Dunod, Paris (2004)
5. Wang, X.-H., Bastidas-Arteaga, E., Gao, Y.: Probabilistic analysis of chloride penetration in reinforced concrete subjected to pre-exposure static and fatigue loading and wetting-drying cycles. Eng. Fail. Anal. **84**, 205–219 (2018)
6. Ragueneau, F., Mazars, J., La Borderie, C.: Damage model for concrete including residual hysteretic loops: application to seismic and dynamic loading. In: Framcos 3 – 3rd Conference on Fracture Mechanics of Concrete Structures, Vol. 1, No. 1, pp. 685–696 1998

Chapter 7
Crack Jumping in Fabric Composite Fracture Testing

Brian T. Werner, Vincente Pericoli, and James W. Foulk III

Abstract Interlaminar fracture is a failure mode that fiber-reinforced polymer composites (FRPC) are commonly susceptible to during loading. The strain energy release rate associated with delaminations can be the limiting factor in a laminate's design. Standard test methods have been developed to measure the critical strain energy release rates using precracked coupons, such as the double cantilever beam (DCB). However, since the adherends in these coupons are laminates themselves, often the crack can initiate a secondary crack within one of the adherends and propagate along a secondary interface as well as the primary, precracked, interface. Deconvoluting the effects of the two cracks, a bridged ply, and multiple crack tips can turn a relatively simple test in determining a material property into a very complicated structural problem. In most cases, it is best to scrap the data collected after the crack has jumped interfaces and start with a fresh specimen. For fabric composites, this phenomenon can be quite common due to the variation in bond line thickness between plies resulting from the architecture of the fabric itself (tow size, weave architecture) as well as manufacturing flaws (voids, foreign object debris). This study aims to use the crack jumping phenomenon to learn more about the characteristics of the process zone through the insertion of designed flaws as well as design a method for evaluating the fracture properties of a toughened film adhesive in a co-cured context.

Keywords Composites · Interlaminar fracture · Cohesive zone modeling · Toughening

7.1 Introduction

Interlaminar failure is a common failure mode in fiber-reinforced polymer composite (FRPC) structures and must be accounted for in any design. Stress concentrations, manufacturing defects, or handling mishaps can lead to initiations of microcracks within the matrix which can grow and result in structural failure. A dominant form of modeling this crack growth is the use of cohesive zone modeling. By measuring the laminate's critical strain energy release rate and assuming a traction-separation law shape, finite element analysis can help predict whether far field stresses will result in crack growth. Interlaminar fracture is typically contained between two plies, but in woven composites, a variation in bond line thickness along with small ply thicknesses can lead to variation in the interlaminar toughness, which can produce crack initiation at other ply interfaces, or crack jumping. The standard method for measuring the Mode I critical strain energy release rate, G_{Ic}, is the double cantilever beam experiment [1]. A major advantage of using this experiment is that the crack length and resulting material resistance can be determined simply from using the load-displacement data and simple beam theory. While high toughness matrix systems can produce significant root rotation at the crack tip which must be accounted for, the relatively low toughness matrix in the composite system used in this study did not have any appreciable beam rotation at the crack tip.

Crack jumping is a phenomenon that can provide significant complications to the analysis of the double cantilever beam (DCB) experiment. Once a secondary, competing, crack has been formed on an adjacent ply, far field stress provides energy to both crack tips and requires much higher loads to advance one of the cracks. In addition to the second crack tip leeching

B. T. Werner (✉) · J. W. Foulk III
Mechanics of Materials Department, Sandia National Laboratories, Livermore, CA, USA
e-mail: btwerne@sandia.gov; jwfoulk@sandia.gov

V. Pericoli
Multiphysics Modeling and Simulation Department, Sandia National Laboratories, Livermore, CA, USA
e-mail: vperico@sandia.gov

S. Xia et al. (eds.), *Fracture, Fatigue, Failure and Damage Evolution, Volume 3*, Conference Proceedings of the Society for Experimental Mechanics Series, https://doi.org/10.1007/978-3-030-60959-7_7

Fig. 7.1 Spaced crack laminate layup; lighter red strips are one ply below the top ply shown

energy from the dominant one, ply bridging is generated by the sandwiched lamina, which can introduce mixed mode loadings at the crack tips and further increase the energy required to advance the cracks. While it is possible to deconvolute all of these phenomena, the complexity is much greater and uncertainty in the resulting material property characterization is usually more of a bother than repeating the experiment. Understanding what conditions may contribute to crack jumping and designing the experiment to mitigate them is a more desirable solution. In this study, a glass fiber reinforced polymer (GFRP) composite laminate was embedded with release film strips, which were placed perpendicular to the direction of crack growth at the main precracked interface and out of phase one ply removed both above and below the main precracked interface. By observing the initiation of the secondary crack, valuable information as to the shape of the traction-separation law can be gleaned for this material. Such information was then used, along with linear elastic fracture mechanics (LEFM), to design a specimen to investigate the Mode I critical strain energy release rate of a GFRP panel that is toughened using a film adhesive.

7.2 Background

The materials used in this study are an 8-harness satin weave GFRP prepreg with a cured ply thickness of approximately 250 microns (0.010 in.). The laminates were embedded with strips of 80micron (0.003 in.) release film in two geometries. Since they were cut and embedded by hand, the two strip sizes were 6.35 mm (0.25 in.) and 12.7 mm (0.5 in.) long; smaller strips were not easy to cut and handle. The strips were spaced 19 mm (0.75 in.) and 37.1 mm (1.5 in.) apart at each interface, and adjacent interfaces were totally out of phase with each other. An image showing the layup in progress is shown in Fig. 7.1.

The cocured panel was designed to investigate the interfacial fracture properties of the base prepreg GFRP, toughened with PL7000 film adhesive, manufactured by Henkel. In this panel, the main precracked interface was toughened as well as the interfaces three plies above and three plies below, as directed by LEFM. Since the main precracked interface was assumed to be three times tougher than the base resin interfaces, if only that interface was toughened, the crack would initiate a secondary crack and jump immediately into one of the adherends. Taking the traction-separation law strength learned from the spaced crack panel, and with the assumption that the stress singularity decays at a rate of $r^{-1/2}$, only three interfaces above and below need to be toughened to ensure that the crack does not jump. The resulting specimen during testing is shown in Fig. 7.2.

Fig. 7.2 Toughened GFRP panel; interfaces near the crack are colored due to presence of PL7000 film adhesive

Fig. 7.3 Crack bridge resulting from crack jumping to an adjacent ply in experiment (left) and cohesive zone modeling simulation (right)

7.3 Analysis

During the spaced crack experiment, the main crack was influenced by the release ply before it fully reached it. The crack then jumped into the adjacent release film and extended along the adjacent ply interface until it neared the release ply on the main crack interface and arrested. The presence of the ply bridge allowed the DCB to load up significantly more before the crack grew forwards from the initial growth plane and backwards from the release ply to eliminate the ply bridge. At this point, the main single crack reformed, was overdriven, and shot forward a significant distance until another bridge stopped it. These results helped determine the shape of the traction-separation law and calibrate the mesh density for the analysis. A comparison between the analysis and experiment is shown in Fig. 7.3.

Initial attempts at a co-cured panel were laid up improperly. The precracked interface is typically made on a plane of symmetry with warp fibers on warp fibers running in the direction of crack growth. For the first attempt, the Teflon precrack was placed on the wrong side, which led to the fibers being perpendicular to the direction of crack growth. This produced a panel where the crack would jump to the weaker warp-weft interface one ply away from the main precracked interface. When the panel was manufactured properly though, the crack was restricted from jumping as designed and the results are shown in Fig. 7.4.

Additional work should be done to ensure that the film adhesive does not produce root rotation as the crack grows, which would explain the positive R-curve response. A properly designed experiment can now lead to proper characterization of toughened composite/film adhesive systems.

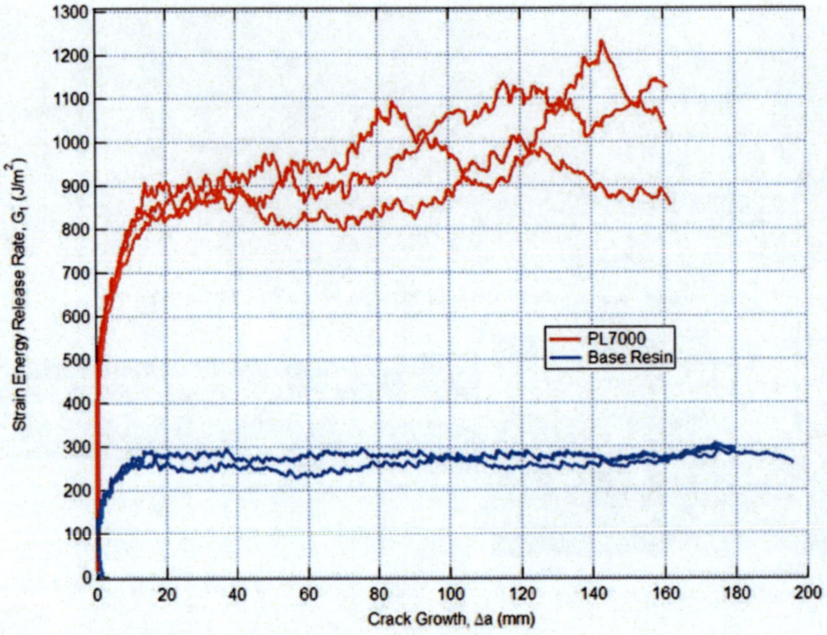

Fig. 7.4 Base GFRP prepreg and co-cured GFRP prepreg with PL7000 film adhesive toughness

Acknowledgment Sandia National Laboratories is a multimission laboratory managed and operated by the National Technology and Engineering Solutions of Sandia, LLC, a wholly owned subsidiary of Honeywell International Inc., for the US Department of Energy's National Nuclear Security Administration under contract DE-NA0003525.

Reference

1. ASTM, ASTM D5528-13: Standard Test Method for Mode I Interlaminar Fracture Toughness of Unidirectional Fiber-Reinforced Polymer Matrix Composites. ASTM International, West Conshohocken (2013)

Chapter 8
Effect of Crack-Parallel Compression or Tension on Mode-I Fracture Energy of Quasibrittle Material – As Applied to Concrete

Hoang Thai Nguyen, Masoud Rezaei, Madura Pathirage, Gianluca Cusatis, Mohsen Issa, and Zdeněk P. Bažant

Abstract In all widely used fracture test specimens, the compressive or tensile stress parallel to the plane of growing crack is negligible, and thus its effect cannot be revealed. The classical fracture models, including the cohesive crack model, cannot capture any effect of such crack-parallel normal stress and strain, except parametrically, because they do not figure such stress and strain as the basic thermodynamic variable. To capture this, the fracture process zone whose 3D stress and strain state is fully described must be implemented. Here it is shown experimentally, and documented by crack band finite element simulations with microplane model M7, that the crack-parallel normal stresses have a major effect on quasibrittle materials such as concrete. They are shown to cause a major decrease or increase of the Mode I (opening) fracture energy G_f (or fracture toughness K_{Ic}). The experiments introduce a modification of the standard three-point bend test, the idea of which is to use plastic pads with a near-perfect yield plateau to first generate compression and a gap at end supports to close later and generate bending. The experiments show and the microplane model confirms that a moderate crack-parallel compression greatly increases G_f (even doubling it), but a higher compression reduces G_f greatly, which represents the case of compression splitting. Through numerical extrapolation, it shows that crack-parallel tension reduces G_f and further that a high compressive or tensile stress normal to the specimen plate has a similar major effect on G_f. While mild parallel stresses arise in shear failure of reinforced concrete beams or slabs and prestressed concrete, high crack-parallel stresses will be impactful in hydraulic fracturing of shale when the effective stress state in the solid phase changes at the presence of a nearby borehole or fluid diffusion.

Keywords Crack-parallel stress · Splitting · Quasibrittle fracture · Microplane model · Crack band theory

8.1 Introduction

In linear elastic fracture mechanics (LEFM), as well as in the cohesive crack model, the normal stress parallel to the crack or notch plane plays no role. Although no such role might be true for fatigue of metals and fine-grained ceramics (except perhaps on the micrometer scale), it is not confirmed by the standard fracture test specimens, because the crack-parallel normal stresses are in all these specimens negligible. The fracture behavior is more complex for quasibrittle materials such as concrete, which are heterogeneous materials made of brittle constituents and possess a fracture process zone (FPZ) whose size is not negligible compared with the structure dimensions. Cubical specimens in homogenous uniaxial compression under lubricated platens fail by an axial splitting crack, with no load applied transversely, as if the fracture energy G_f for Mode I (the opening mode) was zero. This suggests that, in concrete and similar quasibrittle materials, the mode-I G_f should depend on crack-parallel stress σ_{yy} and should terminate with 0 when $\sigma_{yy} \to \sigma_c$ = uniaxial splitting strength. But, again, in

H. T. Nguyen
Theoretical and Applied Mechanics, McCormick School of Engineering, Northwestern University, Evanston, IL, USA
e-mail: hoangnguyen2015@u.northwestern.edu

M. Rezaei · M. Issa
Department of Civil and Materials Engineering, College of Engineering, University of Illinois at Chicago, Chicago, IL, USA
e-mail: mrezae5@uic.edu; missa@uic.edu

M. Pathirage · G. Cusatis · Z. P. Bažant (✉)
Department of Civil & Environmental Engineering, McCormick School of Engineering, Northwestern University, Evanston, IL, USA
e-mail: madurapathirage2014@u.northwestern.edu; g-cusatis@northwestern.edu; z-bazant@northwestern.edu

S. Xia et al. (eds.), *Fracture, Fatigue, Failure and Damage Evolution, Volume 3*, Conference Proceedings of the Society for Experimental Mechanics Series, https://doi.org/10.1007/978-3-030-60959-7_8

the standard fracture test specimens used for concrete, typically the three-point-bend specimen, there is also no significant compression parallel to the notch, and so these tests give no information on its possible effect.

Presented here is a simple modification of the standard notched three-point-bend fracture test that can provide experimental information on the effect of crack-parallel compression. The idea is to apply the compressive load through plastic pads, which exhibit near-perfect plastic yielding with a long yield plateau, and install the end support with a gap that engages only after the compression load has reached the yield plateau. This approach avoids using for compression separate jacks, which is the common way to apply additional loads in structural engineering labs. The self-weight and support of such jacks would make the test evaluation complicated and ambiguous. It will also be shown that the crack band model coupled with microplane constitutive model M7 can match the tests results well. Based on this validation, computer simulations with model M7 will be trusted to extend the understanding of crack-parallel normal stress effects beyond the limited range of the present experiments. A feature of the microplane model that is important for the present problem is that it can capture separately the frictional slip of microcracks of various orientations (modeling internal friction as interaction between the first and second invariant makes no distinction among various orientations of frictional slip).

8.2 Background

Normal concrete of specific compression strength $f'_c = 27.58$ MPa (4000 psi) at 28 days of age was utilized for the experiments. To minimize the scatter of mechanical properties, all the specimens were cast within a few hours from the same batch of concrete delivered by a ready-mix supplier (Ozinga Co.). The specimens for material characterization and model calibration include: cylinders and square prisms for compression tests; cylinders and prisms for splitting tests; beams for fracture tests, geometrically scaled, were of three sizes: small D (height of the notched beam) = 101.6 mm, medium D = 203.2 mm, and large D = 404.6 mm. The elasto-plastic loading pads had sizes S of ratio S/D = 1/4. All the samples were cured for 1 year in a fog room and then tested within 3 weeks. This testing period was sufficiently short to avoid appreciable properties change due to hydration.

To produce additional compressive stress parallel to the crack, elasto-plastic polypropylene pads (Fig. 8.1a) were placed adjacent to the crack mouth (Fig. 8.1a), exactly symmetric with the steel loading pads on the opposite side. Figure 8.1 shows that the pads are almost elastic-perfectly plastic, exhibiting a long yield plateau (strictly speaking there is a very small but negligible hardening slope). A bending moment can then be applied while keeping simultaneous constant compressive stress parallel to crack. To obtain a different value of the applied compressive stress (with no shortening of the yield plateau), holes are drilled through the pads to weaken their yield strength (this also reduces the elastic stiffness). The polypropylene is stiff enough to reach the yield strength without bulging of the pad, and the yielding of the pad results from the growth of shear bands across its body. The complete setup can be seen in Fig. 8.1a (right).

The testing procedure for beams with crack-parallel compression is described in Fig. 8.1b. First, the plastic pads are placed under the beam and the cylindrical supports at beam ends are placed to have a 2–4 mm gap below the bottom surface; Fig. 8.1b-I. Until the gap closes, such elasto-plastic pads produce only the crack-parallel compression with no bending. As the load point displacement increases, the pad deforms first elastically and then yielding begins (Fig. 8.1b-II). Strictly speaking, there is a slight hardening (Fig. 8.1c) due to mild expansion of contact area, but this can be neglected. For a while, the cylindrical end supports are not yet engaged in contact. Once they get in contact, bending begins, and the bending moment M increases at constant crack-parallel compression (Fig. 8.1b-III), which eventually triggers crack growth (Fig. 8.1b-IV). In Fig. 8.1c, this is manifested by an increase of load departing from the yield plateau. This was confirmed to not have any impact on the overall stiffness of the entire beam and fracture analysis but to allow cracks to initiate and propagate at the main crack tip area.

After reaching the peak load, the specimen gets fractured completely, leaving two separate beams under compressive loads. This causes the load to drop down to the yield plateau of the load-displacement curve. From that up and down the portion of the curve, one can extract some of the material response under crack-parallel compression; Fig. 8.1d. A typical load-displacement curve in the whole test is shown in Fig. 8.1c, and the extracted curve of load versus crack tip opening displacement (CTOD) curve is shown in Fig. 8.1d.

8.3 Analysis

To quantify the variation of the fracture energy G_f caused by crack-parallel compression, the size effect method [2] is used. In this method, it suffices to measure the maximum loads, P, of notched specimens of several sufficiently different sizes. Geometrically similar notched three-point-bend beams of three sizes, have been used. To determine G_f, one must obtain first

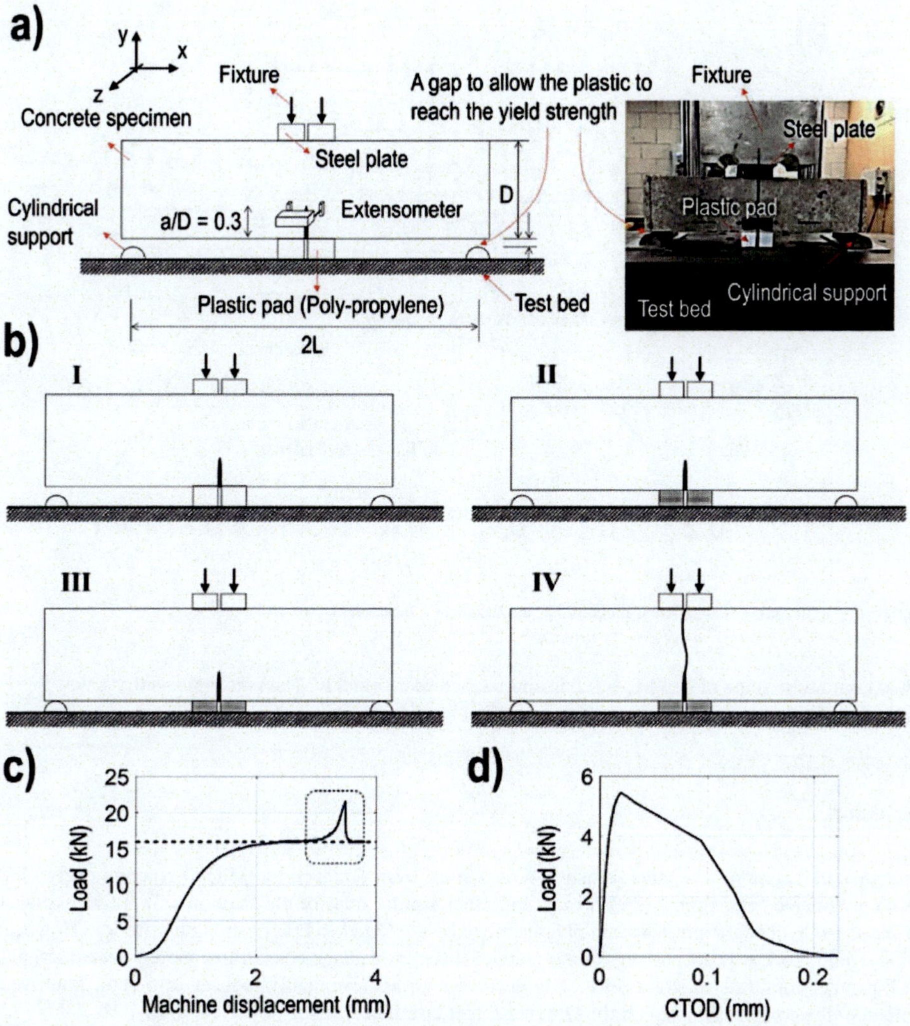

Fig. 8.1 (**a**) Schematic and real experimental configurations; (**b**) testing procedure; (**c**) machine load-displacement responses; and (**d**) the corresponding load – crack tip opening displacement (CTOD) extracted from c

the stress intensity factor, K_I, according to linear elastic fracture mechanics (LEFM) [1, 2]. Although the load configuration is close to three-point bending, four-point-bending with a small but finite distance between the loads must be considered. Note that the larger the size, the greater the effect of crack-parallel compression. The size effect plots of mean log σ_N versus log D are then fitted with the formulation in [2] using the well-known linear least-square regression. The fracture energies G_f and characteristic lengths c_f obtained in this manner for different levels of crack-parallel compression are given as experimental in Fig. 8.2b.

For zero crack-parallel compression, $G_f = 83$ N/m, which is within the range of values reported by many authors [1, 3]. The characteristic length of this case is approximately 18 mm. This is about 1.5 times the average coarse aggregate size. For a moderate crack-parallel compression, $\sigma_{yy} = 0.5\sigma_c$, size effect data yield $G_f = 158$ N/m, which doubles the value at zero compression. The material characteristic length is also nearly doubled, to $c_f = 37.1$ mm, and the FPZ gets more elongated. Another consequence is that the postpeak softening slope is less steep or the occurrence of snapback is suppressed. By contrast, for high crack-parallel compression, $\sigma_{yy} = 0.9\sigma_c$, the fracture energy is drastically reduced, $G_f = 56$ N/m, and so is the value of characteristic length, $c_f = 11.2$ mm, which corresponds to a shorter and wider FPZ, and a reduced c_f. As D to

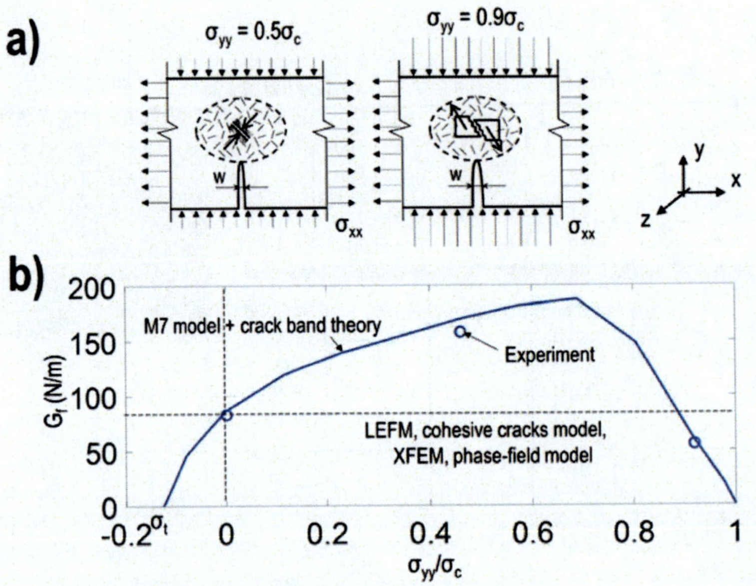

Fig. 8.2 (**a**) Effect of parallel stresses at different magnitudes and (**b**) G_f as a function of σ_{yy}

$\rightarrow \infty$, the final asymptotic slope of LEFM, $-1/2$, is approached more quickly. The mechanisms that cause these phenomena were suggested to increase the volumetric stress on microcracks and expanding slip, see Fig. 8.2a.

8.4 Conclusion

Moderate crack-parallel compressive stress drastically increases, even doubles, the Mode I fracture energy G_f (or fracture toughness K_{Ic}) of concrete, and probably also rock and other similar quasibrittle materials. Intuitive explanation: mainly the effect of increased hydrostatic pressure, simply described by the Drucker-Prager strength criterion. However, the drastic decrease of G_f at high crack-parallel compression requires a different damage constitutive model—the microplane model.

High crack-parallel compressive stress drastically decreases G_f and eventually reduces it to zero. Intuitive explanation: mainly the effect of the compression cap on the Drucker-Pager criterion.

The cohesive crack model, as a line crack model with a scalar relation between crack-bridging stress and relative displacement of crack faces, cannot capture these phenomena, since the crack-parallel stress σ_{yy} and strain are not the basic variables in this model. So σ_{yy} must be used as parameters, which is only an approximation since it cannot capture the effect of the history of triaxial stress and strain.

Linear elastic fracture mechanics, as well as fracture models with scalar damage at the front, which include the scalar phase-field model, cannot capture the effect of crack-parallel compression. To capture the experimentally evidenced effect of crack-parallel compression in general, a tensorial constitutive model for softening damage (which is based on vectorial description at mesoscale cracks) must be used. Most convenient is to use the crack band model coupled with the microplane constitutive law, which can reproduce the inelastic frictional slips separately on planes of different orientations.

An effective method for fracture tests with crack-parallel compression is to generate the compression by plastic pads capable of perfectly plastic yielding and to install the supports with a gap such that they engage in contact only after the pads yield.

The results are of particular interest for the shear failure of reinforced concrete beams and punching of slabs, as well as for the fracture of prestressed concrete, and to hydraulic fracturing of shale, at which the overburden and tectonic stress introduce significant crack-parallel compression.

Acknowledgment Partial preliminary funding under NSF Grant No. CMMI-2029641 to Northwestern University is gratefully acknowledged.

References

1. Bazant, Z.P.: Fracture and Size Effect in Concrete and Other Quasibrittle Materials. Routledge, Boca Raton (2019)
2. Bažant, Z.P., Kazemi, M.T.: Size effect in fracture of ceramics and its use to determine fracture energy and effective process zone length. J. Am. Ceram. Soc. **73**(7), 1841–1853 (1990)
3. Wittmann, F., Mihashi, H., Nomura, N.: Size effect on fracture energy of concrete. Eng. Fract. Mech. **35**(1–3), 107–115 (1990)

Chapter 9
Modal Validation of Academic Bladed Disk with DIC

Dino Celli, Onome Scott-Emuakpor, and Tommy George

Abstract This work is supplementary to extending the use of digital image correlation (DIC) to the experimental method of traveling wave excitation (TWE). The current state of the art for measuring the modal response of a component using TWE is with scanning laser vibrometry (SLV). Although SLV is highly accurate, only one point may be measured during a frequency sweep, therefore potentially causing excessively long experimental testing time when measuring the full-field modal response of a component. The replacement of SLV with DIC has the ability to significantly reduce the time to conduct TWE tests as DIC is capable of measuring full-field data during one sweep when compared to SLV which requires one sweep for often many key points on the component identified by the user. Within this work, DIC measurements are compared and validated against SLV for the measurement of the model response of an academic finger disk undergoing frequency sweeps via single point excitation.

Keywords Digital image correlation · Scanning laser vibrometry · Optical strain measurements · Modal testing · Integrally bladed rotor

9.1 Introduction

Determining the modal response of integrally bladed rotors (IBRs), or blisks (bladed disks), is critical within the aerospace industry. Currently, IBRs are in wide use for modern gas turbine engines due to an increase in performance and decrease in weight when compared to an assembly of rotor and removable blades. However, for the design of gas turbine engines, high-cycle fatigue (HCF) considerations become increasingly more important when utilizing IBRs in actual engine operation as there is no longer inherent damping between the contact of the rotor and inserted blades. Therefore, understanding the harmonic response of an IBR due to force response amplification onset by engine operation or mistuning is very important when determining optimal operating conditions, extending component life, or improving existing maintenance schedules.

Currently, TWE is a simple, cost-effective, and safe method able to determine the modal response of component while simultaneously simulating engine orders all while the part is stationary [1, 2]. Within this procedure, the current state of the art utilizes SLV which is capable of taking highly accurate velocity measurements of components being excited from various excitation sources. However, experimental tests may be exceedingly long because the SLV can only sample one point at a time. It is, therefore, desired to utilize DIC and digital images to gather displacement information as the entire component may be sampled during a single frequency sweep test. For this reason, DIC can significantly reduce the experimental testing time as a SLV must perform each sweep at each designated point of interest. Thus DIC was experimentally validated, by comparing to SLV measurements, to measure the modal response of an academic finger disk undergoing single point excitation.

D. Celli (✉)
Department of Mechanical and Aerospace Engineering, The Ohio State University, Columbus, OH, USA
e-mail: celli.6@osu.edu

O. Scott-Emuakpor · T. George
Air Force Research Laboratory, Wright-Patterson AFB, OH, USA
e-mail: onome.scott-emuakpor.1@us.af.mil; tommy.george@us.af.mil

S. Xia et al. (eds.), *Fracture, Fatigue, Failure and Damage Evolution, Volume 3*, Conference Proceedings of the Society for Experimental Mechanics Series, https://doi.org/10.1007/978-3-030-60959-7_9

Fig. 9.1 Academic finger disk

9.2 Experimental Procedure

First, a reference data set and spectral analysis was collected via SLV or the current measurement approach used within TWE for bladed disks and blisks. The component analyzed in this study was a 16-bladed academic finger disk, shown in Fig. 9.1, and has an outer diameter of 30.48 cm, and each blade has a width and thickness of 1.91 and 0.64 cm, respectively. The academic finger disk was excited via a JODON piezoelectric shaker, and velocity data was taken with a PSV-500 scanning head laser. A sine wave chirp, sweeping through frequencies of 0.5 to 1.5 kHz was generated via Polytec Scanning Vibrometry (PSV) system and hardware. Note, within the PSV software, displacement data was calculated and then exported for direct comparison to DIC data. The frequency resolution for the SLV was 312 mHz and the sample time for each sweep was 3.2 s. There were 177 points of interest defined along the academic finger disk. However, PSV software samples points of interest multiple times, and therefore the experimental testing time was approximately 1 h and 43 min.

For the DIC experimental setup, Photron SA-Z cameras were utilized in stereo to capture time domain displacement data. Note that the Correlated Solutions software was used both for image acquisition and post-processing of images. The working distance of the cameras was 81.28 cm with a stereo angle set to 25 deg. Nikon 50 mm lenses were used to obtain a field of view of 33.02 cm. This yielded a resolution of approximately 0.329 mm/pixels, and for this reason a spray paint speckle pattern was applied with an approximate facet diameter size of 1 mm. A $f\#$ of 8 was found to yield a sufficient depth of field for displacements as well as maintaining focus of entire component. Note, the minimum frame rate required for this test was guided by Nyquist criterion or $(f_{\max}/2)$ or 3000 fps in this case. However, a conservative frame rate of 5000 fps was selected or 1.4 f_{Nyq} with an exposure time of $1/33{,}333$ s. Calibration was performed via a 20 mm Correlated Solutions calibration plate, and a 0.026 calibration score was obtained. Finally, image acquisition was taken over the same period as the sine wave chirp, 3.2 s. Note that for only one scan of the component utilizing DIC requires 3.2 s, whereas the same scan to collect all 177 points of interest using SLV requires $177 * 3.2$ or 566.4 s.

9.3 Experimental Results

In order to obtain mode shape data via DIC, correlation of DIC was first performed in Correlated Solutions VIC-3D with a subset size of 19 and a step size of 3. Eight points spanning the length from the hub to the tip of the top most blade, shown in Fig. 9.1, and out-of-plane displacement data W, along with in-plane spatial coordinates, were sampled for each image. Next an FFT analysis was performed on the displacement data collected via DIC to gather magnitude ($|W|$) and phase (ϕ) data for each individual point. Note, it was assumed the time step between each image was exactly the inverse of the frame rate or 0.2 ms.

For the purpose of validating DIC with SLV measurements, three mode shapes were compared via a modal assurance criterion or MAC. A MAC is a quantitative score ranging from 0 to 1 and is a measure of the similarity between two normalized mode shapes. The equation for MAC can be seen in Eq. (9.1):

$$MAC = \frac{|\sigma_{DIC}^T \sigma_{SLV}|^2}{(\sigma_{DIC}^T \sigma_{DIC})(\sigma_{SLV}^T \sigma_{SLV})} \tag{9.1}$$

where σ_{DIC} and σ_{SLV} are simply normalized vectors of the mode shapes found via DIC and PSV, respectively, and superscript T represents the transpose operator [3].

The frequencies and corresponding MAC scores comparing the similarity between mode shapes found from DIC and SLV are reported in Table 9.1. However, it is important to note that the points taken along the blade by both measurement locations were not precisely at the same location. Therefore the MAC score was calculated by comparing third-order polynomial fits of both DIC and SLV data so that modal response along the length of the blade could be directly compared. The adjusted R^2 terms for each fit is also presented in Table 9.1, and no less than 0.966 which demonstrates each fit is a good approximation of the collected data. The MAC scores presented in Table 9.1 indicates good agreement, or greater than 0.8 MAC score, between DIC and SLV. Additionally in Table 9.1, the maximum of the $(|W|)$ displacements are reported to present a reference to the level of W displacements observed. This is an important note as often TWE measurements are typically within the nano- to micrometer range. For qualitative comparison, the normalized mode shapes of W are presented in Fig. 9.2 and again indicate good agreement between DIC and SLV measurements of mode shape.

9.4 Conclusions

The work presented demonstrates that DIC is an accurate and effective alternative method to laser vibrometry for measuring mode shapes with amplitudes on the order of nanometers with MAC scores greater than 0.95. Additionally, experimental testing time may be reduced by a linear scaling factor equal to the number of points of interest defined when using a SLV.

Table 9.1 MAC score between DIC and SLV

Freq. (Hz)	MAC	max $(\|W\|)$ (nm)	DIC R^2_{adj}	SLV R^2_{adj}
1043.8	0.992	276	0.998	0.985
1131.9	0.979	109	0.986	0.994
1360.6	0.980	228	0.984	0.968

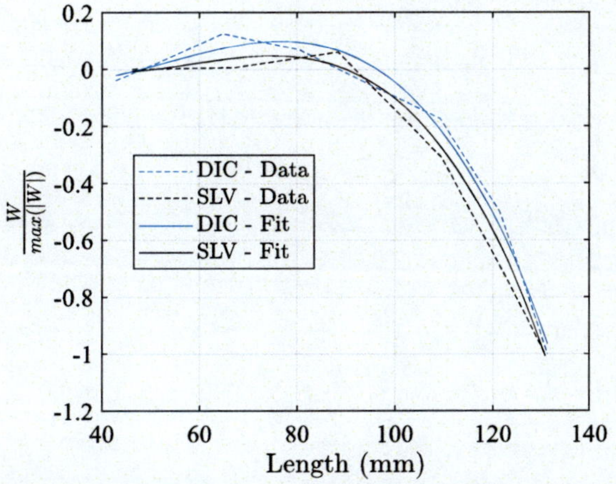

Fig. 9.2 Mode shape comparison of DIC and SLV at 1131.9 Hz

The future work of this paper will initially extend to performing a MAC of the entire academic finger disk as well as an extension to TWE testing. Additionally in future work, an appropriate method must be identified for the application of a speckle pattern for a large-scale component of a highly complex geometry as well as lighting and resolution constraint onset by high-speed image acquisition.

References

1. Feiner, D., Griffn, J., Jones, K., Kenyon, J., Mehmed, O., Kurkov, A.: System identiffication of mistuned bladed disks from traveling wave response measurements. In: International Design Engineering Technical Conferences and Computers and Information in Engineering Conference (2003)
2. Beck, J., Justice, J., Scott-Emuakpor, O., George, T., Brown, J.: Next generation traveling-wave excitation system for integrally bladed rotors. J. Aerosp. Eng. **28**(6), 217–251 (2015)
3. Ewins, D., Modal Testing: Theory and Practice. Wiley, New York, 217–251 (1984)

Chapter 10
Assessing Bond Strength in 304L Stainless Steel Plate Welded Using Plastic Explosives

Thomas A. Ivanoff, Olivia D. Underwood, Jonathan D. Madison, Lisa A. Deibler, and Jeffrey M. Rodelas

Abstract Explosive bonding is a dynamic joining method used to rapidly create metallurgical bonds between two metals. These interfaces can exhibit strengths greater than the parent materials and contain little porosity. Bond quality, however, is highly dependent on processing parameters. Explosive bonds fueled by ammonium nitrate have been extensively characterized, but processing windows for plastic explosives have not. Here, bond strength in 304L stainless steel plate explosively bonded using rubberized/plastic explosives are assessed using shear by tension loading of single-lap-joint specimens. The effects of collision velocity and collision angle on bond quality and strength are investigated and used to define a processing window. Failure modes varied across both the process space studied and within individual bonds. Microstructural analysis across fractured interfaces is combined with fractography to describe the different failure modes and variable strength across bond interfaces.

Keywords Welding · Bond strength · Fractography · 3D characterization · Tension

10.1 Introduction

Explosive bonding is an unconventional metal joining process that utilizes a shock wave to propel a moving object (flyer material) into a fixed object (target material) at sufficient velocity to form a metallurgical bond between the flyer and target materials. A variety of methods can be employed to provide the required shock loading, but detonating explosives are one of the most common [1, 2]. Explosive bonding, or welding, has several advantages over traditional joining methods, such as arc welding or soldering/brazing. First, explosive welding can be used to rapidly form metallurgical bonds between similar and dissimilar metals across large surface areas [3, 4]. Explosive welding is used commercially in large plate and pipe cladding operations [1–3]. The bonds that are formed can be and often are stronger than either base material [5]. Explosive welding, when properly conducted, is a solid-state bonding process. This alleviates the formation of defects such as oxides, remelting products, and shrinkage pores often associated with liquid welding processes.

Explosive welding presents challenges beyond the obvious safety requirements for handling explosives. A main challenge is in reducing nonuniform bonding and ensuring uniform bonds formed across large surface areas do not display varying strengths or levels of porosity. Additional challenges include regions of localized melting and plastic damage to the workpiece. Though these challenges are ever present, proper experimental setup can be used to mitigate most if not all of the challenges associated with explosive bonding.

10.2 Background

In this study, explosive bonding was used as the closure mechanism for a stainless steel gas flow valve in the Z machine at Sandia National Laboratories. Explosive bonding was used because it provided a reliable and fast method of creating a hermetically sealed bond. Hermeticity was necessary to ensure no hazardous compounds escaped from the test environment through the pipe. To evaluate the proper welding conditions for this valve, bonding experiments were completed using plates

T. A. Ivanoff (✉) · O. D. Underwood · J. D. Madison · L. A. Deibler · J. M. Rodelas
Sandia National Laboratories, Albuquerque, NM, USA
e-mail: tivanof@sandia.gov; odunder@sandia.gov; jdmadis@sandia.gov; ldeible@sandia.gov; jmrodel@sandia.gov

S. Xia et al. (eds.), *Fracture, Fatigue, Failure and Damage Evolution, Volume 3*, Conference Proceedings of the Society for Experimental Mechanics Series, https://doi.org/10.1007/978-3-030-60959-7_10

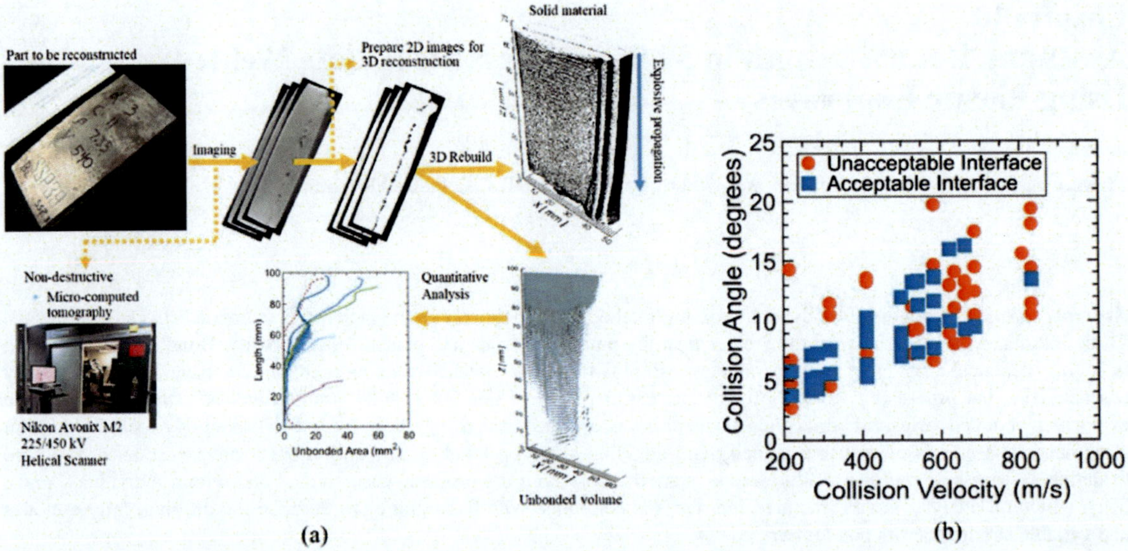

Fig. 10.1 (**a**) Three-dimensional reconstruction procedure for analyzing porosity and unbonded regions within the explosively welded 304L stainless steel plates. Reconstruction allows for quantitative analysis of bond characteristics. (**b**) Process space demonstrating the conditions for successful welds

of 304L stainless steel. Bonding was driven by use of a plastic explosive with a detonation velocity above the speed of sound in steel. It is typical to use slower detonating explosives such as ammonium nitrate and fuel oil (ANFO) for explosive welding, but test constraints in the Z machine precluded this possibility. A series of tests were thus conducted to investigate whether this plastic explosive could be used as a bond driver to provide hermitically sealed bonds with consistent mechanical properties.

Collision velocity and collision angle were identified in the available literature as the two most important bond parameters for producing successful bonds of uniform bonds character [1, 2, 6, 7]. In this study, collision velocity and collision angle were varied to produce bonds with a variety of interfaces. These tests were used to determine the range of parameters that produced successful welds. Similar work has been completed for ANFO and other slow detonating drivers [1, 2]. Successful welds were then characterized using micro-computed tomography and mechanical testing. Three-dimensional reconstructions of the bond interface were digitally created from the micro-computed tomography data and were leveraged to investigate the bond quality. The three-dimensional reconstruction method is illustrated in Fig. 10.1a. These reconstructions allowed for quantitative assessment of the porosity presence in each weld. Three-dimensional reconstructions were created and evaluated using custom processing scripts in IDL, MATLAB, and ImageJ. Bond strength in successful welds was assessed using shear by tension loading of single-lap-joint specimens mechanical testing. Up to four individual test specimens were extracted from each weld and tested.

10.3 Analysis

Successful welds were identified over a range of processing conditions, as demonstrated in Fig. 10.1b. The highest proportion of successful welds formed towards the lower end of the collision velocities observed. Collision angle, however, appears to be less critical to determining bond quality than collision velocity. Three-dimensional reconstructions showed nonuniform bonding along the length of the weld; see "unbonded volume" in Fig. 10.1a. Bonds often contained waves of porosity that varied in size along the bond length. The degree and intensity of this porosity varied with collision velocity and some bonds demonstrated little to no observable porosity.

Significant variability in mechanical performance was observed both (1) between welds produced using different welding parameters and (2) between specimens tested from within the same bond. The variability between welds produced using different welding parameters is likely related to the change in porosity presence between bonds. It is suspected that the primary reasons for variability within a given bond are likely inconsistent bond quality, i.e. porosity presence along the interface, and variability introduced during testing. Failure in the tension specimens was observed both along the interface and within the parent material associated with the grip regions. Fractography revealed that different failure modes (tension, shear, and bending) were dominant in different failure locations. This is likely a result of variable bond strength within specimens and bond interfaces hardened relative to the grip material.

10.4 Conclusion

Successful welds can be produced in 304L stainless steel plate using plastic explosives with detonation velocities exceeding the speed of sound in steel, but the processing parameter window is small. Significant variability in porosity presence and strength were observed along bonded interfaces. This inconsistency in bond quality produces welds with highly variable interface strength, even though they may demonstrate hermiticity.

Acknowledgment Sandia National Laboratories is a multimission laboratory managed and operated by the National Technology & Engineering Solutions of Sandia, LLC, a wholly owned subsidiary of Honeywell International Inc., for the US Department of Energy's National Nuclear Security Administration under contract DE-NA0003525.

This chapter describes objective technical results and analysis. Any subjective views or opinions that might be expressed in the chapter do not necessarily represent the views of the US Department of Energy or the United States Government. SAND2020-2227 A.

References

1. Crossland, B.: Explosive Welding of Metals and its Application. Oxford University Press, Oxford (1982)
2. Blazynski, T.Z.: Explosive welding, forming and compaction. Appl. Sci. (1983)
3. Zamani, E., Liaghat, G.H.: Explosive welding of stainless steel – carbon steel coaxial pipes. J. Mater. Sci. **47**, 685–695 (2012)
4. Athar, M.M.H., Tolaminejad, B.: Weldability window and the effect of Interface morphology on the properties of Al/Cu/Al laminated composites fabricated by explosive welding. Mater. Des. **86**, 516–525 (2015)
5. Crossland, B.: The development of explosive welding and its application in engineering. Met. Mater., 401–413 (1971)
6. Crossland, B., McKee, F.A., Szecket, A.: An experimental investigation of explosive welding parameters. In: High-Pressure Science and Technology, pp. 1837–1845. Springer US, New York (1979)
7. Ribeiro, J.B., Mendes, R., Loureiro, A.: Review of the weldability window concept and equations for explosive welding. J. Phys. **500**, 1–6 (2014)

Chapter 11
Real-Time Visualization of Damage Progression Inside GFRCs via High-Speed X-Ray PCI Technique

Jinling Gao, Nesredin Kedir, Cody Kirk, Julio Hernandez, Xuedong Zhai, Junyu Wang, Tyler Tallman, Kamel Fezzaa, and Weinong Chen

Abstract Insight into damage progression within glass fiber reinforced composites (GFRCs) contributes to understanding failure of composites by interaction of various damage modes, developing physics-based canonical theoretical models, and finally manufacturing desired compositions. In this work, dynamic singe-edge notched bending (DSENB) experiments were performed on a modified Kolsky compression bar, impacting the notched composite beam onto an indenter mounted in front of a load cell. The high-speed X-ray phase-contrast imaging (PCI) technique was used to penetrate the opaque composite and capture in real time the detailed damage initiation and evolution inside the GFRC. Experimental results were compared with those obtained by optical imaging technique, revealing high-speed X-ray PCI technique was able to characterize the inner layers of composite and capture the damage progression among multiple composite layers.

Keywords GFRC · Dynamic loading · Damage evolution · High-speed X-ray PCI technique

11.1 Introduction

GFRC has wide applications as structural materials in helmets, vehicle armor, truck box, and military vehicle hulls. This material possesses excellent impact resistance compared with the composites reinforced by other engineering fibers, such as Kevlar®, Dyneema®, and Zylon. Simultaneously, it has higher toughness than carbon fiber reinforced composites [1–2]. One major drawback of such composite is its susceptibility to damage caused by foreign object impact, which induces debonding of individual plies, stiffness degradation, and finally catastrophic failure of the complete composite structure [2].

Current damage assessment techniques for fiber reinforced composites (FRCs) are mainly through insertion of a sensor inside the composite specimen [3] or non-destructive techniques [4–7]. When a small-diameter sensor is embedded into the composite, additional defects might be introduced which will easily alter the original attributes of cracks or damage the composite being investigated. Acoustic emission [5] is an effective method for inspection analysis on the composites, using an array of highly sensitive piezoelectrics to detect the stress waves spreading from the defects inside the materials, such as matrix microcracking, fiber-matrix debonding, localized delamination, or fiber pullout and breakage. However, such technique has difficulty in providing quantitative crack size information. On the other hand, radiographic inspection techniques [6–7] such as X-rays are able to visualize the internal features within solid objects and obtain digital information for damage tracking.

J. Gao (✉) · C. Kirk · J. Hernandez · X. Zhai · J. Wang · T. Tallman
School of Aeronautics and Astronautics, Purdue University, West Lafayette, IN, USA

N. Kedir
School of Materials Engineering, Purdue University, West Lafayette, IN, USA
e-mail: nkedir@purdue.edu; wchen@purdue.edu

K. Fezzaa
Advanced Photon Source, Argonne National Laboratory, Argonne, IL, USA

W. Chen
School of Aeronautics and Astronautics, Purdue University, West Lafayette, IN, USA

School of Materials Engineering, Purdue University, West Lafayette, IN, USA

S. Xia et al. (eds.), *Fracture, Fatigue, Failure and Damage Evolution, Volume 3*, Conference Proceedings of the Society for Experimental Mechanics Series, https://doi.org/10.1007/978-3-030-60959-7_11

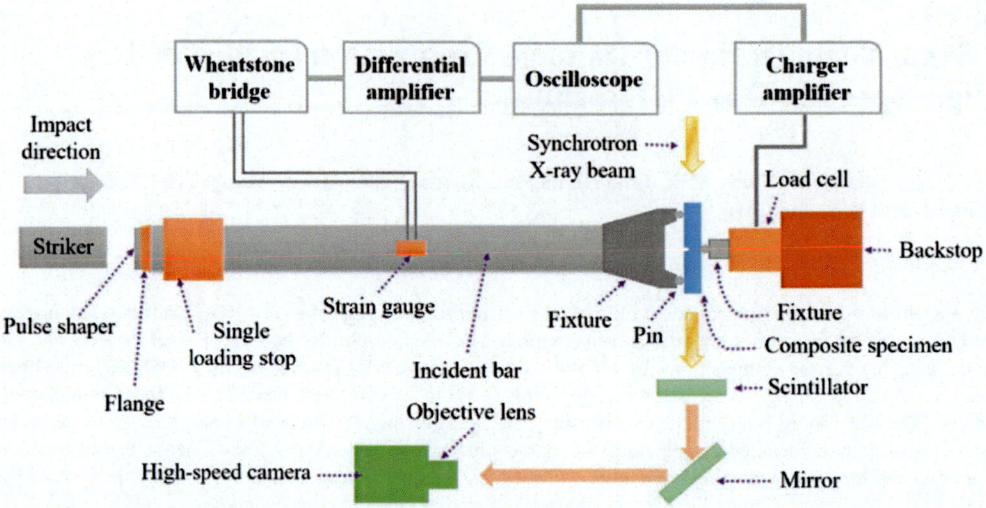

Fig. 11.1 Schematic of the experimental setup

In this work, we transversely loaded a pre-notched S-2 glass/SC-15 composite beam at a speed of ~6 m/s on a modified Kolsky compression bar. We visualized in real time the damage evolution inside the composite in front of the notch by virtue of high-speed X-ray phase contrast imaging technique. The experimental results were then compared with the damage progression captured by the optical imaging technique.

11.2 Materials and Methods

S-2 glass fiber roving was provided by AGY World Headquarters and manufactured into unidirectional fabric. Each fabric was then brushed with SC-15 matrix (Applied Poleramic, Benicia, CA, USA) and then laid up to 24 layers. Composites were cured in a vacuum bag at room temperature for 24 h and given another 24 h for additional cure. Post-curing was conducted at 120 °C for 2 h without vacuum. The large panel was then cut by a diamond saw into smaller beams with a span (*l*) of ~18 mm and a width (*b*) of ~2.5 mm, referring to ASTM D5045-14 [8]. The height (*h*) of the beam was ~5 mm.

The experimental setup is schematically described in Fig. 11.1. A notched composite beam specimen was mounted at the bar end through a designed fixture. The beam contacted with the fixture at two cylinders with a distance of 15 mm. During the experiment, the bar was impacted by a striker, driving the specimen to the indenter fixed ahead of a load cell. The indenter transversely loaded the specimen until the material failed. The synchrotron X-ray penetrated the specimen from the side and captured the crack geometry in real time, which was transferred as the optical signal by a scintillator, then magnified via a 5X optical lens, and finally sent to a high-speed camera for recording.

11.3 Results and Discussions

Figures 11.2 and 11.3 compare dynamic failure of the pre-notched GFRC beam captured via the high-speed X-ray PCI technique and common optical imaging technique. Both of the two composites were stacked in the manner of $[0°/90°]_{12}$. Obviously, X-ray penetrated the specimen and enabled to identify individual layers of the composite, while the optical method was only able to observe the surface morphology of the specimen. Once loading, the high-speed X-ray PCI technique captured in accuracy the notch opening, matrix cracking, fiber bridging, fiber/matrix debonding, fiber breakage, and delamination around the notch tip inside the composite. In comparison, the optical imaging technique can merely recognize roughly the opened notch and delamination around the notch tip. However, due to the limited size of the beamline, damage evolution inside the composite was only observed within an area of 2.56 mm × 1.6 mm around the notch tip. In contrast, by changing the objective lens, the optical imaging technique was capable of revealing the delamination away from the notch tip.

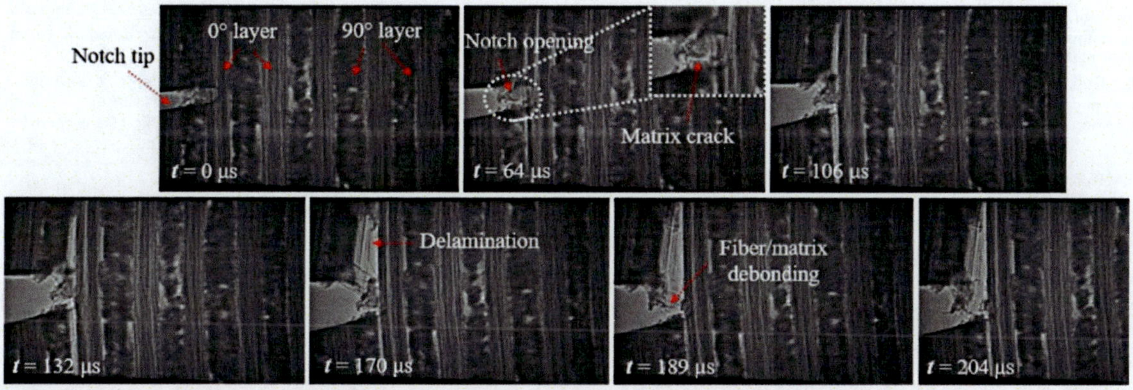

Fig. 11.2 Dynamic failure behavior of the GFRP composite beam captured by the high-speed X-ray PCI technique

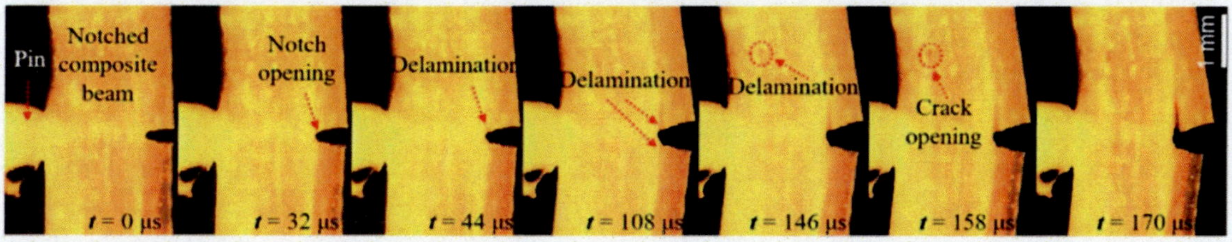

Fig. 11.3 Dynamic failure behavior of the GFRP composite beam captured by the optical imaging technique

11.4 Conclusion

This work presents the dynamic failure behaviors of pre-notched GFRC beams subjected to low-velocity impact via the high-speed X-ray PCI technique. By comparing with the experimental results via common optical imaging technique, it is revealed that the high-speed X-ray PCI technique is able to capture accurately the damage evolution inside the composites, including the matrix cracking, fiber bridging, fiber pullout, fiber/matrix debonding, fiber breakage, and delamination. Such decent damage detection enables to further understand the physical damage progression inside the composites and calculate the fracture toughness in quantification.

Acknowledgments This research was sponsored by the Army Research Laboratory and was accomplished under Cooperative Agreement Number W911NF-12-2-0022. The views and conclusions contained in this document are those of the authors and should not be interpreted as representing the official policies, either expressed or implied, of the Army Research Laboratory or the US Government. The US Government is authorized to reproduce and distribute reprints for government purposes notwithstanding any copyright notation herein.

References

1. Kulkarni, S.G., Gao, X.L., Horner, S.E., Zheng, J.Q., David, N.V.: Ballistic helmets–their design, materials, and performance against traumatic brain injury. *Compos. Struct.* **101**, 313–331 (2013)
2. Sathishkumar, T.P., Satheeshkumar, S., Naveen, J.: Glass fiber-reinforced polymer composites–a review. *J. Reinf. Plast. Compos.* **33**(13), 1258–1275 (2014)
3. Takeda, S., Minakuchi, S., Okabe, Y., Takeda, N.: Delamination monitoring of laminated composites subjected to low-velocity impact using small-diameter FBG sensors. *Compos. A: Appl. Sci. Manuf.* **36**(7), 903–908 (2005)
4. Fard, M.Y., Sadat, S.M., Raji, B.B., Chattopadhyay, A.: Damage characterization of surface and sub-surface defects in stitch-bonded biaxial carbon/epoxy composites. *Compos. Part B.* **56**, 821–829 (2014)
5. Gholizadeh, S., Leman, Z., Baharudin, B.T.H.T.: A review of the application of acoustic emission technique in engineering. *Struct. Eng. Mech. Int. J.* **54**(6), 1075–1095 (2015)

6. Bull, D.J., Spearing, S.M., Sinclair, I., Helfen, L.: Three-dimensional assessment of low velocity impact damage in particle toughened composite laminates using micro-focus X-ray computed tomography and synchrotron radiation laminography. *Compos. A: Appl. Sci. Manuf.* **52**, 62–69 (2013)
7. Bull, D.J., Spearing, S.M., Sinclair, I.: Investigation of the response to low velocity impact and quasi-static indentation loading of particle-toughened carbon-fibre composite materials. *Compos. A: Appl. Sci. Manuf.* **74**, 38–46 (2015)
8. D5045–14: Standard test methods for plane-strain fracture toughness and strain energy release rate of plastic materials 1. *ASTM J.* **99**, 1–9 (2014)

Chapter 12
Watching High-Cycle Fatigue with Automated Scanning Electron Microscope Experiments

Nathan M. Heckman, Timothy A. Furnish, Christopher M. Barr, Khalid Hattar, and Brad L. Boyce

Abstract Fatigue is a multistep process where cyclic loading causes damage within materials that eventually leads to crack formation and propagation. In nanocrystalline metals, a dominant damage mechanism is the abnormal growth of grains up to 100 times their original size. Previous in situ synchrotron experiments have revealed that this grain growth process precedes crack formation and takes up a majority of the fatigue lifetime. The growth of nanocrystalline grains leads to the formation of protrusions on the surface of a material, which can be resolved in scanning electron microscopy. Based on this concept, an automated in situ scanning electron microscope tension–tension fatigue test method has been developed to observe the evolution of crack formation and propagation in materials. In this study, this method was applied to understand the high-cycle fatigue behavior in nanocrystalline Ni- and Pt-based alloys. Fatigue tests between 10^5 and 10^7 cycles were performed, and in combination with postmortem characterization through grain orientation mapping and transmission electron microscopy, we identified differences in resistance to damage and crack propagation in the various alloys, and observed varying damage levels prior to crack formation, strongly dependent on the number of cycles to failure.

Keywords Fatigue · In situ · Scanning electron microscopy · Crack initiation · Crack propagation

12.1 Introduction

Fatigue is often the cause for mechanical failure of components, where a vast majority of mechanical service failures in metals are due to cyclic loading events [1–4]. While this is the case, most studies investigate uniaxial monotonic properties of materials instead of fatigue properties. Due to this, the monotonic tensile deformation behavior of simple and complex face centered cubic (FCC) metals are well understood across a wide range of microstructures, where dislocation motion tends to dominate when grain sizes are in the coarse-grained regime [5], while at the nanocrystalline scale, grain-boundary-mediated mechanisms tend to dominate the monotonic deformation behavior [6]. Monotonic deformation behavior is typically well understood even in advanced metals across these grain sizes. There is a much lesser understanding of complex mechanisms in fatigue. In this manuscript, we discuss a newly developed method to understand the evolution of fatigue damage in metals utilizing automated in situ fatigue experiments in the scanning electron microscope (SEM). This method is utilized in preliminary studies on nanocrystalline Ni-based alloys to understand the evolution of fatigue damage in these material systems. This serves as a tool to understand both simple and complex mechanisms in the formation and propagation of damage in fatigue.

12.2 Background

The two primary stages of fatigue failure are crack formation and crack propagation. Each stage is often mechanistically dissimilar, where different mechanisms may dominate during each. In comparison to coarse-grained metals, nanocrystalline metals tend to spend a much larger portion of their fatigue life in the crack formation stage. In most simple nanocrystalline systems, abnormal grain growth has been observed as the dominant mechanism for crack formation [7]. Note, this mechanism

N. M. Heckman (✉) · T. A. Furnish · C. M. Barr · K. Hattar · B. L. Boyce
Sandia National Laboratories, Albuquerque, NM, USA
e-mail: nheckma@sandia.gov; tafurni@sandia.gov; cbarr@sandia.gov; khattar@sandia.gov; blboyce@sandia.gov

S. Xia et al. (eds.), *Fracture, Fatigue, Failure and Damage Evolution, Volume 3*, Conference Proceedings of the Society for Experimental Mechanics Series, https://doi.org/10.1007/978-3-030-60959-7_12

is not universal for all nanocrystalline metals, however, where, for example in nanocrystalline CuAl containing a high percentage of coherent twin boundaries, detwinning was observed as the dominant fatigue crack initiation mechanism, followed by intergranular fracture between twinned and detwinned columnar grains [8].

A previous study has demonstrated the capability to observe abnormal grain growth throughout the fatigue life of nanocrystalline Ni-Fe utilizing an x-ray diffraction technique in a synchrotron on a notched specimen [9]; in this study, a test was interrupted prior to complete crack initiation, where it was observed in SEM imaging that a clear protrusion formed along the fracture surface that corresponded with the abnormally enlarged grain. This serves as the motivation for the technique in this study, where in situ fatigue tests were performed inside of a JEOL IT300HR SEM. An in situ piezoelectric tensile stage described in previous experiments [10] was utilized for these experiments. Because of the long duration of fatigue tests and short duration of the crack propagation regime in nanocrystalline metals, the SEM image acquisition was automated so that images were acquired throughout the entire fatigue life of the sample to capture the full progression of crack initiation and propagation.

A demonstration of this technique was performed on a nanocrystalline Ni-Fe film with a grain size of 50 nm. A 5-micron thick film was synthesized by electrodeposition and cut into dog-bone geometries utilizing a femtosecond laser system. A notch was cut into the sample using an FEI Nanolab dual beam focused ion beam (FIB), following procedures outlined in a previous study [10]. Each sample was loaded in tension–tension load control using a stress ratio (minimum to maximum stress), R, of 0.3. Fatigue tests were performed with a sinusoidal wave with a frequency of 30 Hz, and fatigue tests were interrupted every 1000 cycles to image the sample in the SEM. Figure 12.1 illustrates the test setup as well as several SEM images from a Ni-Fe sample loaded with a stress amplitude of roughly 200 MPa that failed at a total of 2.25×10^5 cycles.

12.3 Analysis

In the example presented in Fig. 12.1, a clear morphological feature forms along the sample's notch at 2.00×10^5 cycles. This feature is consistent with features observed previously in nanocrystalline Ni-Fe which corresponded to an abnormally enlarged grain [9]. Note that even though the feature was only observed at 2.00×10^5 cycles, power law growth is expected in this system, and unresolved protrusions may have formed prior to this observation. This feature reaches a maximum size of less than 1 micron prior to crack initiation, after which a crack forms and propagates. The crack propagates with a steady crack growth rate of roughly 7×10^{-9} m/cyc in the images shown at a ΔK of roughly 5.0 MPa$\sqrt{m}$. The crack then propagates to failure at a total of 2.25×10^5 cycles.

12.4 Conclusion

The technique allows for observation of the full evolution of fatigue damage throughout the fatigue life of a material system. Coupled with techniques such as microstructural characterization through electron backscatter diffraction (EBSD) and postmortem fractography, one can characterize not only how fatigue damage propagates throughout cyclic loading in materials but also the correlation between mechanisms of fatigue crack initiation/propagation and the fatigue strengths of the material.

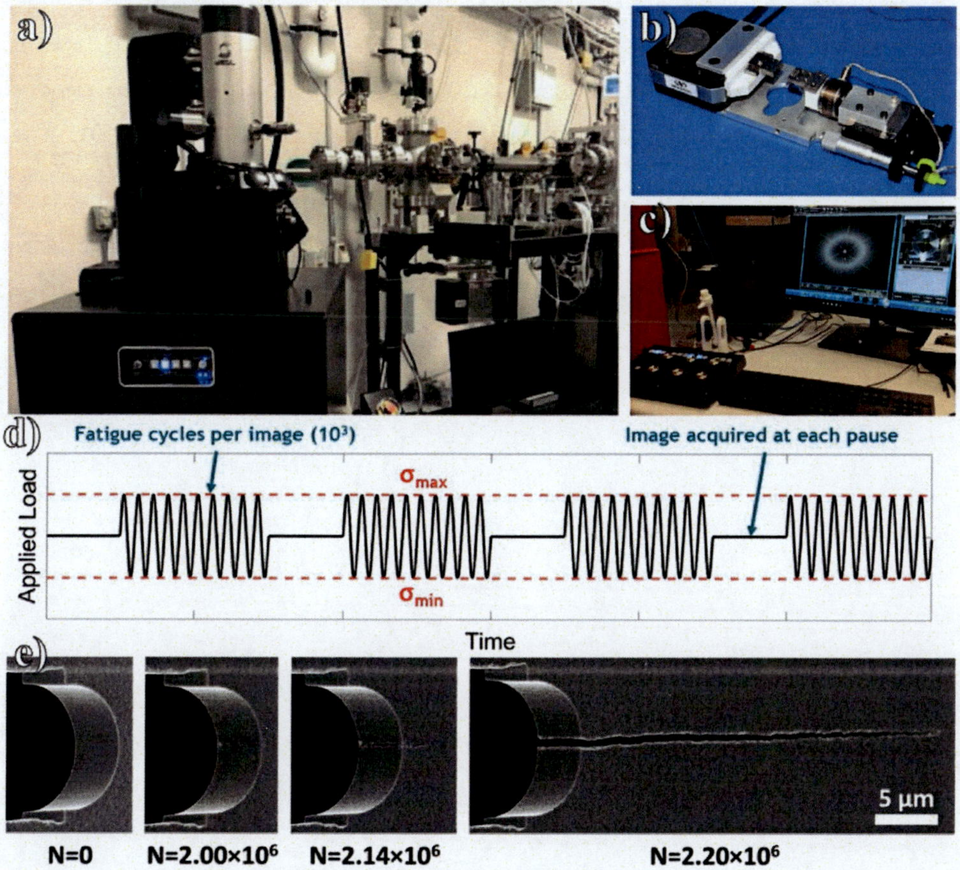

Fig. 12.1 Example of in situ SEM fatigue test performed on nanocrystalline Ni-Fe. The tests were performed inside of a JEOL IT300HR SEM (**a**) which housed a custom piezoelectric fatigue stage (**b**). Images were acquitted automatically in the SEM utilizing a custom Basic Image Recording Device (**c**). Fatigue loads were applied according to the schematic in (**d**) where 1000 fatigue cycles were applied at fixed loads, and then the loading was paused at the mean load to acquire an image. The progression of crack propagation in nanocrystalline Ni-Fe can be observed at different total fatigue cycles, *N* (**e**)

Acknowledgments This work was funded by the United States Department of Energy (DOE), Office of Basic Energy Sciences (BES) (Grant No. 15013170), Division of Materials Science and Engineering. The authors acknowledge use of microscopy facilities at the Center for Integrated Nanotechnologies, an Office of Science User Facility operated for the US DOE Office of Science by Sandia National Laboratories. Sandia National Laboratories is a multi-mission laboratory managed and operated by National Technology and Engineering Solutions of Sandia, LLC., a wholly owned subsidiary of Honeywell International, Inc., for the US Department of Energy's National Nuclear Security Administration under contract DE-NA0003525. Any subjective views or opinions that might have been expressed in the chapter do not necessarily represent the views of the US Department of Energy or the United States Government.

References

1. Campbell, F.C.: Elements of Metallurgy and Engineering Alloys. ASM International, Materials Park (2008)
2. Gurney, T.R.: Fatigue of Welded Structures. CUP Archive, Cambridge (1979)
3. Hosford, W.F.: Mechanical Behavior of Materials. Cambridge University Press, Cambridge (2010)
4. Stephens, R.I., Fatemi, A., Stephens, R.R., Fuchs, H.O.: Metal Fatigue in Engineering. Wiley, New York (2000)
5. Hertzberg, R.W., Hauser, F.E.: Deformation and fracture mechanics of engineering materials. J. Eng. Mater. Technol. **99**(1), 96 (1977)

6. Kumar, K.S., Van Swygenhoven, H., Suresh, S.: Mechanical behavior of nanocrystalline metals and alloys. Acta Mater. **51**, 5743–5774 (2003)
7. Padilla, H.A., Boyce, B.L.: A review of fatigue behavior in nanocrystalline metals. Exp. Mech. **50**, 5–23 (2010)
8. Heckman, N.M., Berwind, M.F., Eberl, C., Hodge, A.M.: Microstructural deformation in fatigued nanotwinned copper alloys. Acta Mater. **144**, 138–144 (2018)
9. Furnish, T.A., Bufford, D.C., Ren, F., Mehta, A., Hattar, K., Boyce, B.L.: Evidence that abnormal grain growth precedes fatigue crack initiation in nanocrystalline Ni-Fe. Scr. Mater. **143**, 15–19 (2018)
10. Heckman, N.M., Foiles, S.M., O'Brien, C.J., Chandross, M., Barr, C.M., Argibay, N., Hattar, K., Lu, P., Adams, D.P., Boyce, B.L.: New nanoscale toughening mechanisms mitigate embrittlement in binary nanocrystalline alloys. Nanoscale. **10**, 21231–21243 (2018)

Chapter 13
Determination of Mixed-Mode (I/III) Fracture of Polycarbonate

Ali F. Fahem, Vijendra Gupta, Addis Kidane, and Michael A. Sutton

Abstract In many engineering applications, Mode-III type loading in the crack tip region is more common. Since loading in structures oftentimes is quite complex, the crack tip region generally experiences Mixed-Mode conditions. In this work, torsional loading experiments are performed by employing a modified spirally cracked cylindrical specimen. The cylindrical specimen used in all experiments is machined to incorporate a full revolution, spiral v-notch crack. The v-notch crack is inclined at an angle of 67.5° with respect to the specimen centerline to obtain Mixed-Mode (I/III) crack tip conditions under torsional loading. By combining the experimental measurements with detailed numerical simulations, the Mixed-Mode (I/III) fracture parameters for polycarbonate (PC) are quantified using an interaction integral method. The elastic-viscoplastic material response of the PC material, required for numerical simulations, is determined by performing standard tensile loading experiments. The Mixed-Mode (I/III) fracture toughness, as well as the stress intensity factors for Mode-I and Mode-III crack tip conditions are presented and discussed.

Keywords Spiral crack · Mixed-Mode (I/III) fracture · Torsion load · Numerical method · Polycarbonate

13.1 Introduction

Engineering applications of amorphous polycarbonate (PC) material have expanded in the last 20 years. The general mechanical behavior of this type of material is elastic-viscoplastic, which is inelastic and rate dependent [1]. Thus, understanding its behavior in both static and dynamic conditions is important for structural integrity and safe design. In fracture mechanics, the notch sensitivity, crazing phenomena, yield strength, ductile-brittle fracture transition, and fracture toughness were studied by many investigators in both static [2–6] and dynamic [1, 7] cases. However, most of these studies used the loading conditions of a single fracture mode, i.e. Mode-I, Mode-II, or Mode-III, and few of them tested Mixed-Mode (I/II) fracture [8]. Internal crazes and plastic deformation ahead of a crack tip were the most commonly observed behavior in PC materials. The craze condition, which appears to be a type of plastic deformation, is referred to as brittle fracture in the polymer [9]. Liu et al. studied the Mixed-Mode (I/III) of fracture mechanics of Polymethyl Methacrylate plastic (PMMA) and Al. 7050. They used a circumferentially notched cylindrical bar subjected to two far-field loading conditions, torsion and uniaxial tension [10]. They investigated the brittle and ductile crack behavior transition related to the fracture mode. They found that "the elastic theory predicts the crack propagation direction and the tensile-shear transition accurately for both brittle and ductile engineering materials." Boyce, in 1986 [11], studied the rate-dependent constitutive model of glassy polymers.

In this work, torsional far-field loading was applied on a cylindrical spirally cracked specimen to generate stress fields related to a Mixed-Mode condition around the crack front, including Mode-I, Mode-III, and Mixed-Mode (I/III). The fracture toughness for Mixed-Mode (I/III) conditions was calculated numerically by using the experimental data as input. The experimental data used in the simulations included the material's behavior (elastic-viscoplastic) and the fracture load. Unlike

A. F. Fahem (✉)
Department of Mechanical Engineering, University of Al-Qadisiyah, Al-Qadisiyah, Iraq

Department of Mechanical Engineering, University of South Carolina, Columbia, SC, USA
e-mail: Ali.Fahem@qu.edu.iq; afahem@email.sc.edu

V. Gupta · A. Kidane · M. A. Sutton
Department of Mechanical Engineering, University of South Carolina, Columbia, SC, USA
e-mail: vijendra@email.sc.edu; kidanea@cec.sc.edu; sutton@sc.edu

S. Xia et al. (eds.), *Fracture, Fatigue, Failure and Damage Evolution, Volume 3*, Conference Proceedings of the Society for Experimental Mechanics Series, https://doi.org/10.1007/978-3-030-60959-7_13

the traditional methods, the spiral crack specimens were used to measure Mode-I fracture toughness of the static and dynamic conditions as well [12–15]. Two types of specimens were tested. Dumbbell-shaped specimens are tested in uniaxial tension while spiral v-notch spiral crack specimens are used to determine the Mixed-Mode (I/III) fracture toughness. The uniaxial tension test provides the PC material e\response used in finite element modeling to calculate fracture toughness parameters for PC.

13.2 Determination of Material Behavior of PC

13.2.1 Specimen Preparation and Experimental Setup

Polycarbonate is known to be an elastic-viscoplastic material. Dumbbell-shaped specimens of Type IV, as shown in Fig. 13.1, are used in uniaxial tension experiments, as per American Society for Testing and Materials (ASTM) D638–14 [16]. The specimen is prepared by machining to the size shown in Fig. 13.1. The experiments are performed with a loading rate of 1 mm/min. The load is acquired from the load cell mounted on the load frame, while the strain is measured using 2D-DIC (Digital Image Correlation). The in-plane strain components are acquired using Vic-2D software, a 2D-DIC software from Correlated Solutions Inc. [17, 18].

13.2.2 Uniaxial Tension Response

The engineering stress versus strain curve is shown in Fig. 13.2. The load versus deformation was recorded up to failure. In general, PC shows quasi-ductile behavior with elastic and viscoplastic conditions. The elastic modulus and Poisson's ratio of the material were calculated using the linear elastic part of the stress-strain curve and are mentioned in Table 13.1. The material yielded at 63.7 MPa.

The true stress versus strain curve can be obtained from the engineering stress-strain curve and is shown in Fig. 13.3. The behavior of the PC can be divided into four stages, as shown in Fig. 13.3. *Stage one* starts from point O and ends at point A. This stage has a linear elastic behavior with initial hardening developing near point, the initial yield stress of the material. Segment AB represents *Stage two*. In this region, the stress drops and it is also referred to as the softening stage. *Stage three,* from point B to point C, is the section with maximum extension. In this stage the material exhibits viscoplastic behavior (orientation hardening). Point C corresponds to the ultimate tensile stress. *Stage four* is the last stage where the stress drops by small amount below the ultimate tensile stress. This stage ends at the failure denoted by point D. Thus, it is seen that PC exhibits quasi-ductile behavior with elastic and viscoplastic behavior.

The plastic strain was obtained by decomposing the strain into elastic and plastic strain components. The true stress-plastic strain data along with the material properties shown in Table 13.1 were input into the finite element model. Abaqus, Finite Element Analysis (FEA) was used for all the numerical simulations in this study. A numerical simulation was performed on a uniaxial test specimen (model without crack is shown in Fig. 13.4a) to verify that the finite element output gave the same

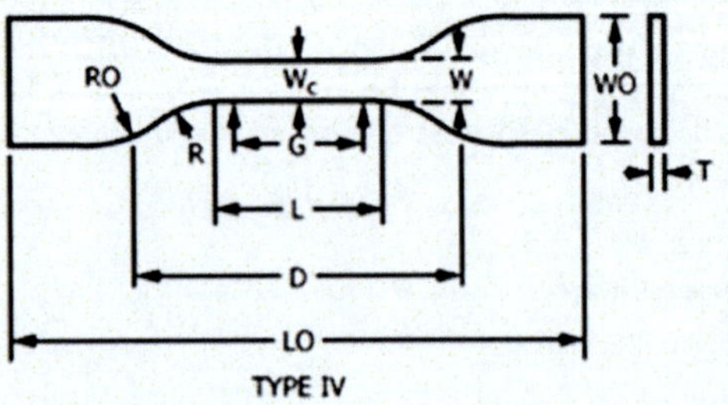

- W=Wc: Width of reduced section = 6 mm
- L: length of narrow section = 33 mm
- WO: Width overall = 19 mm
- LO: length overall = 115 mm
- G: Gage length = 25 mm
- D: Distance between grips = 65 mm
- R: radius of fillet = 14 mm
- RO: Outer radius = 25 mm
- T: Thickness = 3.8 mm

Fig. 13.1 Dumbbell-shaped specimen (ASTM D638–14, Type IV)

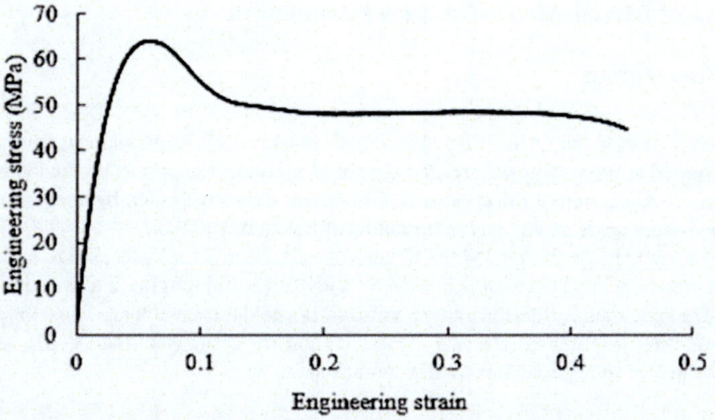

Fig. 13.2 Typical engineering stress-strain response of PC under uniaxial tension using ASTM D638–14

Table 13.1 Material properties of PC

Elastic modulus (GPa)	Poisson's ratio	Yield stress (MPa)
2.41	0.418	63.7

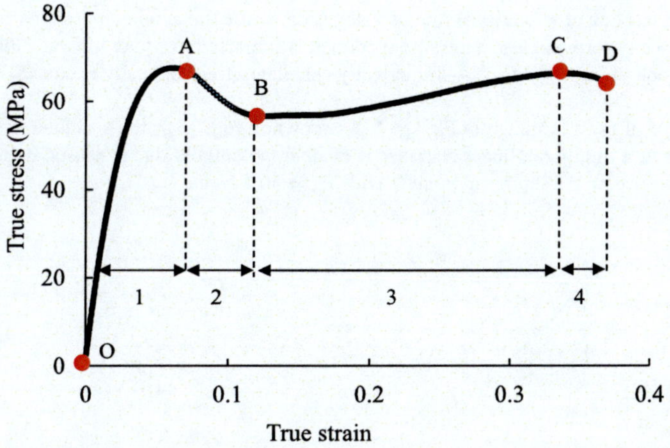

Fig. 13.3 True stress—true strain response of PC under uniaxial tension

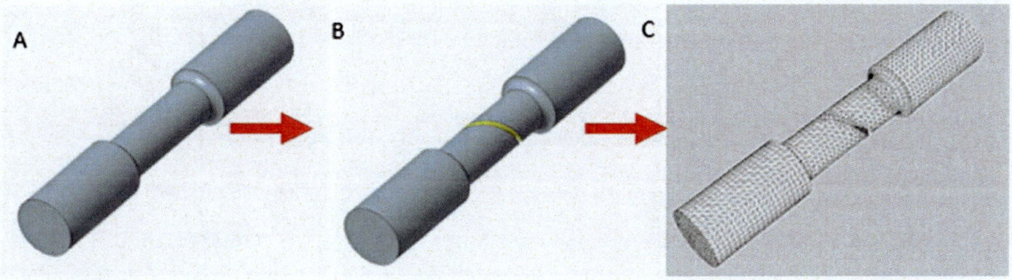

Fig. 13.4 Finite element model

input material response. Then, the spiral crack was added, and the finite element model with mesh is shown in Fig. 13.4b, c. Details regarding the finite element model and the boundary conditions can be found in previous work [19, 20].

13.3 Determination of Mixed-Mode Fracture Parameters

13.3.1 Specimen Preparation

Clear impact-resistant polycarbonate bar (ASTM D3935-PC) of radius $r = 12.7$ mm sourced from McMaster-Carr are used. All experiments are performed at room temperature. A fully spiral v-notch crack is cut on the cylindrical surface of the bar, as shown in Fig. 13.5. The cross-section of the specimen, Fig. 13.5b, shows the crack ligament of $a = 8.7$ mm, crack depth of 4 mm, and a crack tip v-notch angle of 45° with a tip radius of less than 100 μm.

A stainless-steel small curving blade (item #35435A67 on McMaster-Carr.com) was used to make the final sharp artificial crack. The spiral v-notch has an inclined angle of $\beta_{sp} = 67.5°$ with the central longitudinal axis of the bar for all experiments performed in this study. The specimen was tested under a uniform far-field torsional load. Thus, the crack front was subjected to two components of stress, uniaxial tensile stress σ_n (Mode-I) and shear stress σ_t (Mode-III), as shown in Fig. 13.5c. In general, the crack tip is subjected to Mixed-Mode (I/III) conditions.

13.3.2 Fracture Load Measurement

A tension-torsion machine manufactured by Test Resources is used for all experiments. Unless the experiment is carefully designed, a compressive load could develop during loading since the specimen extends during twisting. The spiral v-notch specimen is twisted up to fracture at a rotational rate of 1 deg./min, while the axial compression load is controlled and kept equal to zero throughout the experiment test. Thus, the specimen is subjected to a pure torsional load condition. The fracture torque obtained from the measurement is used to numerically calculate Stress Intensity Factor (SIF) for the cylindrical solid specimen (Fig. 13.6).

The torque versus angle of twist is shown in Fig. 13.7. As shown in Fig. 13.7, the specimen response is linear elastic up to point A. Beyond point A, a lightly nonlinear response is evident, potentially due to plastic deformation around the crack front, continuing until the specimen fractures at point B with $T_B = 40.1$ N-m.

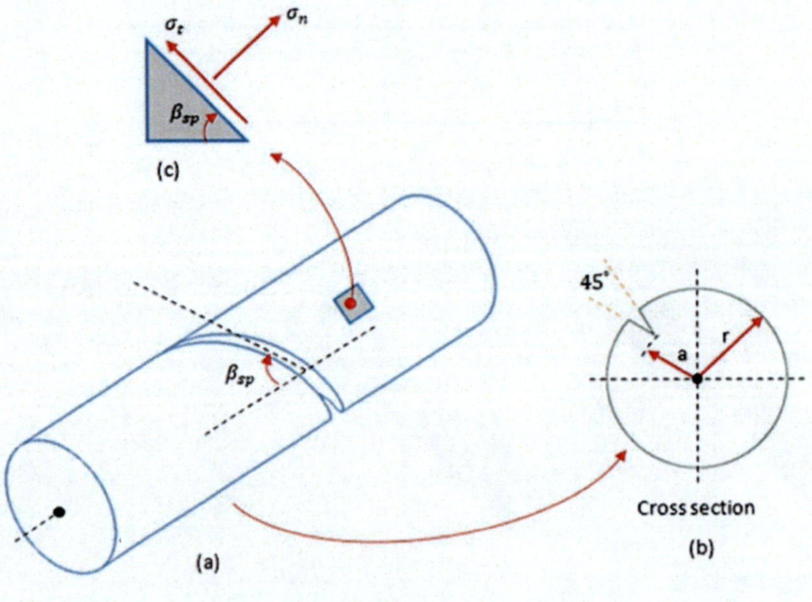

Fig. 13.5 Schematic of PC specimen with spiral crack and effective stresses

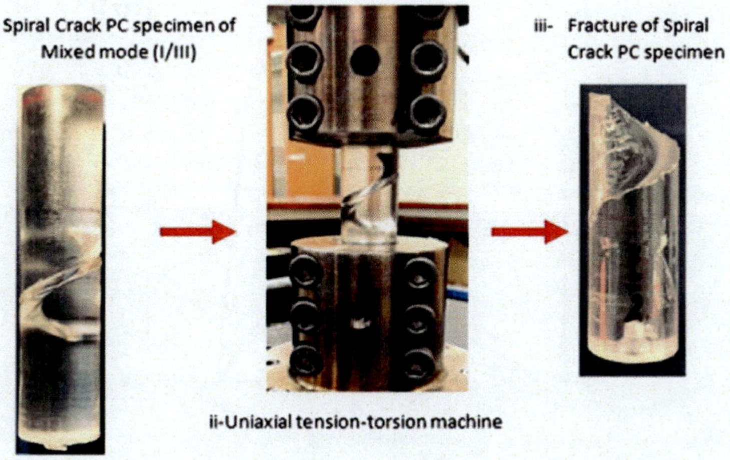

Fig. 13.6 Mixed-Mode fracture experimental setup using PC specimen in a uniaxial tension-torsion machine, β = 67.5°

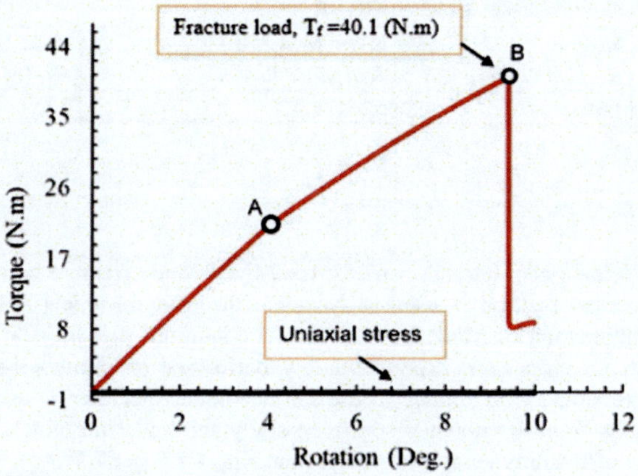

Fig. 13.7 Torque vs rotation angle for a spiral v-notch specimen up to fracture, β = 67.5°

13.3.3 Numerical Computation of SIFs and J-Integral

The fracture load is used as input to the numerical model for calculation of the J-integral and the stress intensity factors for Mode-I, Mode-III and Mixed-Mode (I/III) loading conditions. The theoretical solution of the spirally notched specimen using a 3D interacting integral method is available in previous work [21, 22]. As shown in Fig. 13.8, there is sufficient fracture load to generate brittle fracture at the notch front with a small jump just prior to fracture initiation moment. In general, fracture occurs under nominally plane strain conditions. From Fig. 13.8 and Table 13.2, the fracture toughness of the PC specimen under Mixed-Mode conditions indicates that $K_{Ic} > K_{IIIc}$ for $\beta = 67.5°$, with the in-plane Mode-II component near zero [23]. Finally, the total Mixed-Mode (I/III) fracture toughness of PC, given by Eq. (13.1), is higher than the fracture toughness of Mode-I, i.e. $K_{(I/III)c} > K_{Ic}$:

$$K_{I/IIIc} = \sqrt{K_I^2 + \frac{K_{III}^2}{1 - \nu}} \tag{13.1}$$

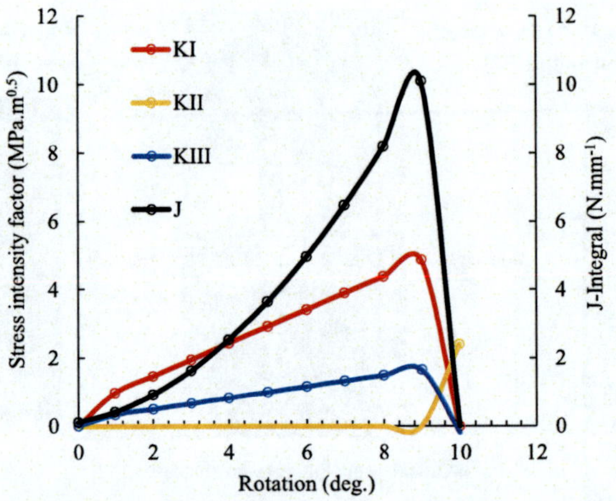

Fig. 13.8 Stress intensity factor of spirally notched PC specimen for β = 67.5°

Table 13.2 Fracture toughness result of Mixed-Mode PC (static case), β = 67.5°

Material	Mode-I $K_I\ (MPa\sqrt{m})$	Mode-III $K_{III}\ (MPa\sqrt{m})$	Mixed-Mode(I/III) $K_{I/IIIc}\ (MPa\sqrt{m})$	Mode-I [4] $K_{Ic}\ (MPa\sqrt{m})$	J-Integral (J/m^2)
PC	4.85	1.65	5.264	4.1	10.1

13.4 Conclusion

For a specific spiral v-notch specimen geometry, β = 67.5°, the Mixed-Mode fracture behavior of PC material is studied. A combined experimental-numerical method is used to calculate the Mixed-Mode (I/III) fracture toughness of elastic viscoplastic material for the combination of Mode I and Mode III conditions present near the notch tip when β = 67.5°. To obtain the baseline PC material properties, experiments are performed on dumbbell-shaped tension specimens. The experimental data for both the torsional load applied to the spiral v-notch specimen is used as input to calculate fracture toughness of the material. Results from the notched specimen study show that the notch tip is undergoing Mixed-Mode conditions, with KIc at the onset of fracture nearly 3X greater than K_{IIIc} for β = 67.5°.

Acknowledgments The support provided by the Ministry of Higher Education and Scientific Research, University of Al-Qadisiyah, College of Engineering, Mechanical Engineering Department, Iraq is greatly acknowledged. The support of the Department of Mechanical Engineering at the University of South Carolina and the Center for Mechanics, Materials, and NDE in the development and maintenance of the dynamic experimental equipment is deeply appreciated. The support of Dr. Jordan at the Air Force Office of Scientific Research and the Department of Defense DURIP program is gratefully appreciated.

References

1. Faye, A., Basu, S., Parameswaran, V.: Effect of loading rate on dynamic fracture toghness of polycarbonate. Dyn. Behav. Mater., 139–145 (2014)
2. Cho, K., Yang, J.H., Kang, B.I., Park, C.E.: Notch sensitivity of polycarbonate and toughened polycarbonate. J. Appl. Polym. Sci. **89**, 3115–3121 (2003). https://doi.org/10.1002/app.12502
3. Gearing, B.P., Anand, L.: Notch-sensitive fracture of polycarbonate. Int. J. Solids Struct. **41**, 827–845 (2004). https://doi.org/10.1016/j.ijsolstr.2003.09.058
4. Salazar, A., Rodríguez, J., Martinez, A.B.: Fracture toughness reliability in polycarbonate: notch sharpening effects. Indian J. Mater. Sci. **2013**, 1–4 (2013). https://doi.org/10.1155/2013/187802
5. LD, G.: Crazing, yielding, and fracture of polymers. I. Ductile brittle transition in polycarbonate. J. Appl. Polym. Sci. **13**, 2129–2147 (2018)
6. Parvin, M., Williams, J.G.: Ductile-brittle fracture transitions in polycarbonate. Int. J. Fract. **11**, 963–972 (1975). https://doi.org/10.1007/BF00033842
7. Sundaram, B.M., Tippur, H.V.: Dynamic mixed-mode fracture behaviors of PMMA and polycarbonate. Eng. Fract. Mech. **176**, 186–212 (2017)

8. Ravi-Chandar, K.: On the failure mode transitions in polycarbonate under dynamic mixed-mode loading. Int. J. Solids Struct. **32**, 925–938 (1995)
9. Ishikawa, M., Noarisawa, I., Ogawa, H.: Criterion for craze nucleation in polycarbonate. J. Polym. Sci. Polym. Phys. Ed. **15**, 1791–1804 (1977)
10. Liu, S., Chao, Y.J., Zhu, X.: Tensile-shear transition in mixed mode I/III fracture. Int. J. Solids Struct. **41**, 6147–6172 (2004)
11. Boyce, M.C.: Large Inelastic Deformation of Glassy Polymers. PhD thesis, Massachusetts Institute of Technology (1986)
12. Kidane, A., Wang, J.: A new method for dynamic fracture toughness determination using torsion Hopkinson pressure bar. In: Dynamic Behavior of Materials, Conference Proceedings of the Society for Experimental Mechanics, pp. 307–312, New York (2013)
13. Wang, J.-A., Liu, K.C.: Fracture Toughness Determination Using Spiral-Grooved Cylindrical Specimen and Pure Torsional Loading, p. 2 (2003)
14. Fahem, A., Kidane, A.: Modification of Benthem solution for mode I fracture of cylinder with spiral crack subjected to torsion. Fract. Fatigue Fail. Damage Evol. Proc. Soc. Exp. Mech. Ser. **6**, 57–63 (2019). https://doi.org/10.1007/978-3-319-95879-8_10
15. Fahem, A., Kidane, A.: A progression on the determination of dynamic fracture initiation toughness using spiral crack. Fract. Fatigue Fail. Damage Evol. Conf. Proc. Soc. Exp. Mech. Ser. **6**, 89–95 (2019). https://doi.org/10.1007/978-3-319-95879-8_15
16. ASTM: Designation: ASTM D638-14 standard test method for tensile properties of plastics. ASTM Int. (2015). https://doi.org/10.1520/D0638-14
17. Sutton, M.A., Orteu, J.J., Schreier, H.W.: Image Correlation for Shape, Motion and Deformation Measurements- Basic Concepts, Theory and Applications. Image Rochester NY 341 (2009)
18. Sutton, M.A., Deng, X., Liu, J., Yang, L.: Determination of elastic-plastic stresses and strains from measured surface strain data. Exp. Mech., 99–112 (1996)
19. Fahem, A., Kidane, A.: Hybrid computational and experimental approach to identify the dynamic initiation fracture toughness at high loading rate. Dyn. Behav. Mater. Conf. Proc. Soc. Exp. Mech. **1**, 141–146 (2018). https://doi.org/10.1007/978-3-319-62956-8_24
20. Fahem, A., Kidane, A.: A general approach to evaluate the dynamic fracture toughness of materials. Dyn. Behav. Mater. Conf. Proc. Soc. Exp. Appl. Mech. **1**, 185–194 (2017). https://doi.org/10.1007/978-3-319-41132-3_26
21. Fahem, A., Kidane, A., Sutton, M.: Mode-I dynamic fracture initiation toughness using torsion load. Eng. Fract. Mech. **213**, 53–71 (2019). https://doi.org/10.1016/j.engfracmech.2019.03.039
22. Fahem, A.F., Kidane, A., Sutton, M.: A model for calculating geometry factors for mode I stress intensity factor of a cylindrical specimen with spiral crack subjected to torsion. Eng. Fract. Mech. under review (2019)
23. Beer, F.P., Johnston, E.R., DeWolf, J.T.: Mechanics of Materials. McGraw-Hill, Boston (2006)

Chapter 14
Influence of Dynamic Multiaxial Transverse Loading on Dyneema® SK76 Single Fiber Failure

Frank David Thomas, Stephen L. Alexander, C. Allan Gunnarsson, Tusit Weerasooriya, and Subramani Sockalingam

Abstract The primary objective of this research is to investigate, through fundamental experiments, the dynamic multiaxial deformation, failure and strength degradation mechanisms that govern individual ballistic fiber failure. Predicting ballistic impact performance of armor systems requires an understanding of fiber failure under complex multiaxial loading conditions. This study examines the failure behavior of ultrahigh molecular weight polyethylene (UHMWPE) Dyneema® SK76 single fibers under dynamic transverse impact as a function of varying loading rates and projectile geometry. A novel single fiber transverse impact experiment is developed by modifying the Kolsky bar to characterize failure of fibers to create the foundation for a failure model. Experiments are performed with load cells at the fiber ends and with high speed imaging for determining average stresses and strains. Post-test microscopy imaging of failure surfaces are compared to determine the dominant fiber failure modes for each experimental group.

Keywords UHMWPE · Ballistic impact · Transverse compression · High strain rate

14.1 Introduction

Polymeric fibers such as ultra-high molecular weight polyethylene (UHMWPE) are selected for ballistic applications based on their high specific stiffness and specific strength [1, 2]. Characterization of tensile properties of UHMWPE-based material systems have been performed at both quasi-static and high-rate speeds for varying length scales, ranging from single Dyneema® SK76 fibers to full woven fabrics [3, 4]. The knowledge gained through these experiments is incorporated into computer models through the multiscale mechanics of materials approach, which in turn improves predictive modeling capabilities.

For ballistic performance predictions, an understanding of fiber behavior under multiaxial loading is essential. A diverse array of experiments have been performed to quantify various components of the material response of ballistic fiber to multiaxial loading, including quasi-static transverse loading of yarns and individual fibers [5], high strain rate transverse compression and measurement of single filament residual strength at quasi-static and high strain rates [6, 7], and high loading rate transverse impact of yarns [8, 9]. The quasi-static multiaxial loading has been shown to reduce the fiber tensile strength [5–7]. Experimental data combined with representative models have been applied in the development of a strain-based failure criterion for single fiber multiaxial loading at varying strain rates. However, characterizing the transverse loading behavior of individual single fibers at high strain rates is challenging. This chapter details the development of an experimental technique to characterize the failure of single fibers subjected to multiaxial loading as a function of loading rate and geometry.

F. D. Thomas (✉) · S. Sockalingam
McNAIR Center for Aerospace Innovation and Research, University of South Carolina, Columbia, SC, USA

Department of Mechanical Engineering, University of South Carolina, Columbia, SC, USA
e-mail: fthomas@email.sc.edu; SOCKALIN@mailbox.sc.edu

S. L. Alexander
SURVICE Engineering Company, Belcamp, MD, USA
e-mail: stephen.l.alexander18.ctr@mail.mil

C. A. Gunnarsson · T. Weerasooriya
US Army Research Laboratory, Aberdeen Proving Ground, MD, USA
e-mail: carey.a.gunnarsson.civ@mail.mil; tusit.weerasooriya.civ@mail.mil

S. Xia et al. (eds.), *Fracture, Fatigue, Failure and Damage Evolution, Volume 3*, Conference Proceedings of the Society for Experimental Mechanics Series, https://doi.org/10.1007/978-3-030-60959-7_14

14.2 Experimental Setup

An experimental technique is developed by modifying the 0.25 inch diameter Kolsky bar to conduct single fiber transverse impact, as shown in Fig. 14.1. The indenter is attached to the incident bar end through a sleeve, and a U frame is used to secure the load cells (Kistler 9712B5) attached to the single fiber mounted on a window frame. A Photron Fastcam SA-5 high speed camera is used to record the deformation. Projectile-fiber contact induces multiaxial loading and strain concentration in the fibers. The setup allows measuring load at failure, strain at failure, failure angle θ, and impact velocity. A series of experiments are performed to characterize fiber failure under dynamic multiaxial loading conditions and to better understand the influence of different projectile geometry (blunt to sharp to razor) on fiber failure.

Diameter measurements are performed on each sample using a confocal optical microscope in a region of the fiber which is proximal to the intended impact location and averaged for use in appropriate stress calculations based on load measurements. As a result, failure at the impact site is a necessary criterion for an experiment to be considered valid in this study.

The fibers are impacted at velocities in the range of 10–20 m/s. Measurements made with camera recordings utilize the overall diameter of the projectile, which is 1 mm, for scale. In order to record at a higher framerate (100,000 fps), resolution is reduced to 320×192. The location of the impactor tip is measured in the initial impact frame and in the frame before failure, and the distance between the two points is divided by the elapsed time to yield the average velocity. A pressure vessel is used to control the velocity, which is filled to specific high and low pressures to vary velocity. Outlier velocities for a given pressure setting are classified as irregular test conditions and are therefore excluded from this study.

Analytical calculations are used to determine strains and strain rates from observed geometric changes. The maximum angle of deflection before failure (θ) provides a useful means of approximating far-field strains (ε) according to Eq. 14.1. Strain rates under the indenter are difficult to evaluate based on the available data, so a hybrid finite element approach must be used to quantify the relationship between strain rate at the failure location and impact velocity. However, for discussion purposes in this chapter, the approximate strain rate (*SR*) is related to velocity *v* according to Eq. 14.2, where gage length L_O is 41.5 mm. The piezoelectric load cells to which the fiber ends are mounted measure axial loads in the fiber at a high frequency and can be used to estimate average stress in the fiber.

$$\varepsilon = \frac{1}{\cos\theta} - 1 \tag{14.1}$$

$$SR = \frac{v}{\frac{1}{2}L_O} \tag{14.2}$$

In order to examine the effects of high-rate multiaxial loading on single UHMWPE fibers, several indenters with varying loading geometries have been designed and manufactured through wire EDM. Loading geometry in all cases is circular and altered by changing the radius in orders of magnitude relative to the fiber diameter. The average fiber diameter of Dyneema® SK76 is measured to be 17.7 μm. The sharp indenter has a radius on the same order of magnitude as the fiber diameter

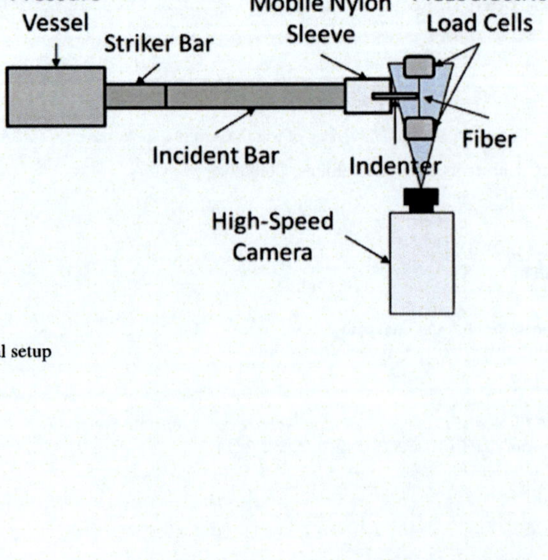

Fig. 14.1 Schematic of experimental setup

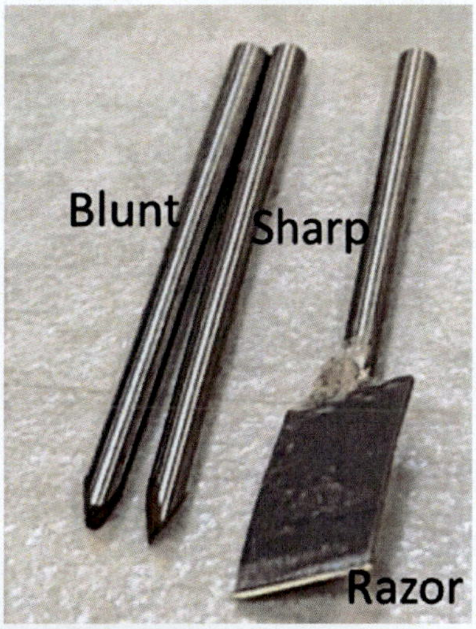

Fig. 14.2 Indenter geometry comparison

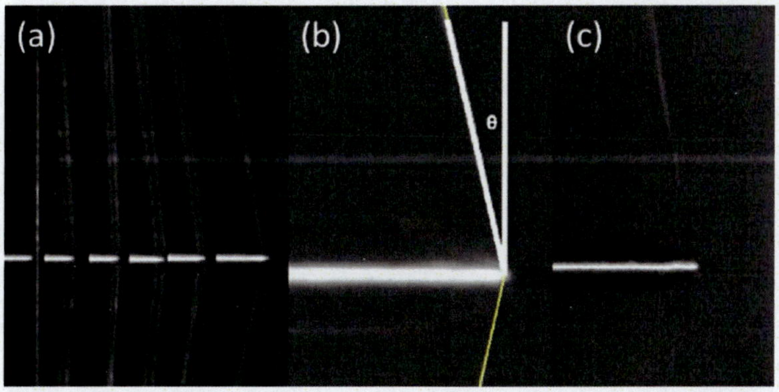

Fig. 14.3 Experimental image analysis. (**a**) Progressive loading of single fiber. Images range from undeformed state (far left) to final ultimate tensile strain before failure (far right). (**b**) Angle measurement in final frame before failure. (**c**) Fiber motion post-failure

at approximately 20.0 μm. The blunt indenter is one order of magnitude larger with a radius of 200.0 μm, and the razor indenter is one order smaller at approximately 2.0 μm radius. The indenters as produced are depicted in Fig. 14.2. For each loading geometry, a minimum of 10 successful tests at each loading velocity were performed.

14.3 Results and Discussion

Tests are split into 6 groups based on a combination of impact velocity (fast or slow) and indenter geometry (blunt, sharp, or razor). Each test group contains a minimum of 10 samples in each. Figure 14.3 represents a typical test progression as used for measuring velocity and failure strain. Fast tests correspond to velocities of approximately 20 m/s with a pressure set point of 15 psi, and slow tests correspond to an impacting velocity of approximately 10 m/s with a pressure set point of 5 psi. Average velocities for all groups with standard deviation are displayed in Fig. 14.4.

Sanborn et al. have performed tensile testing of Dyneema® SK76 single fibers at varying gage lengths and strain rates, the results of which are included in Fig. 14.5 [3]. Based on this study, as strain rate increases in uniaxial tension, failure strain decreases, but apparent strength increases, presumably due to rate-dependent stiffness increase. Average experimental strain

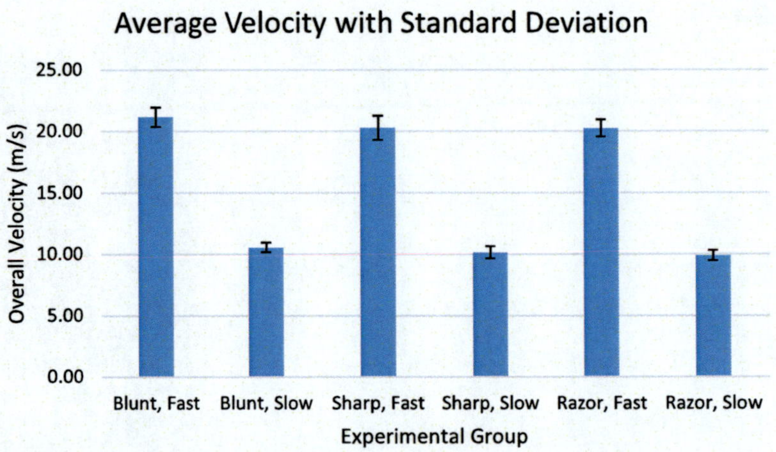

Fig. 14.4 Test velocity comparison

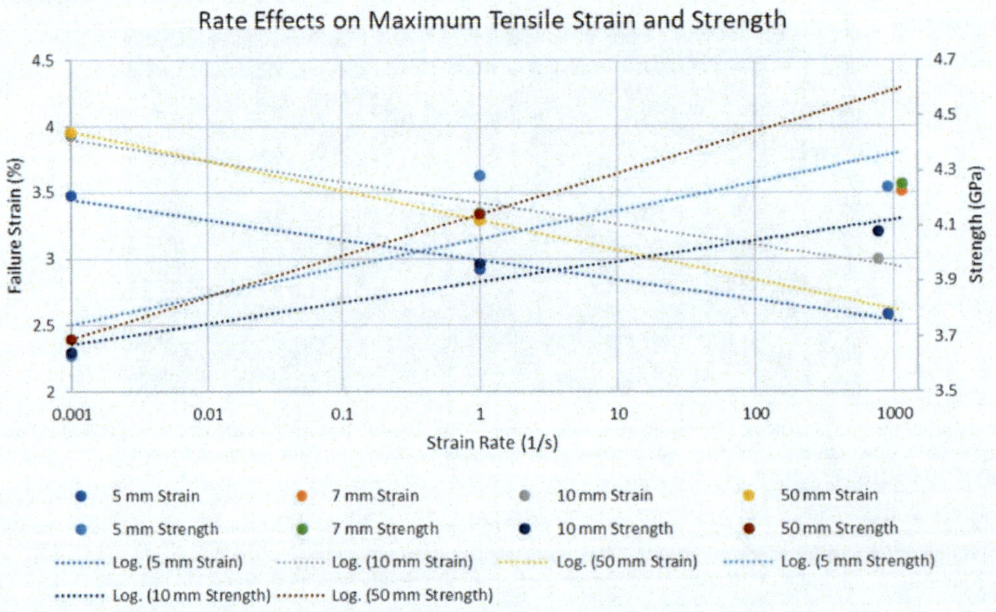

Fig. 14.5 Summary of Sanborn et al. Failure strain decreases with increasing strain rate while apparent strength increases [3]

values calculated using Eq. 14.1 are displayed in Fig. 14.6 with comparison to comparable tensile data. Strength values are calculated based on the average of the maximum loads recorded by each load cell before fiber failure divided by the average diameter at the point of impact. Representative load curves for each group are displayed in Fig. 14.7. Average stress values and standard deviations are displayed in Fig. 14.8 with comparable tensile data. Sample sizes and graph data are summarized in Table 14.1. Controlling for geometry, both strength and failure strain reduce with increase in nominal strain rate.

Broken fiber ends were imaged via light microscopy, and representative images are displayed in Fig. 14.9. Visible damage varies largely according to loading geometry but not so much with average strain rate, with blunt indenters inducing highly visible fibrillation, sharp indenters producing shorter fibrillation (as indicated by the greater magnification on those images), as well as some fibril shearing, and razor indenters producing little to no fibrillation.

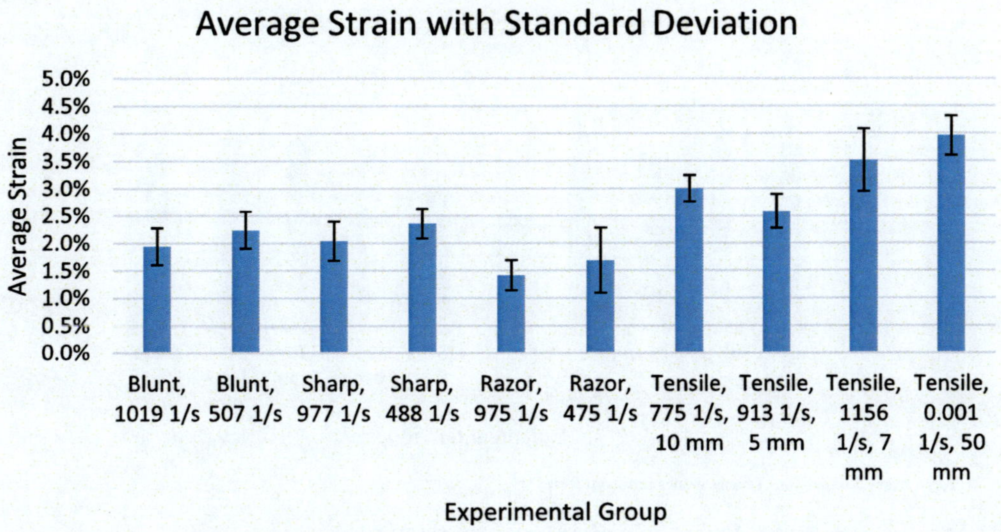

Fig. 14.6 Average strain comparison. Tensile values are from Ref. [3]

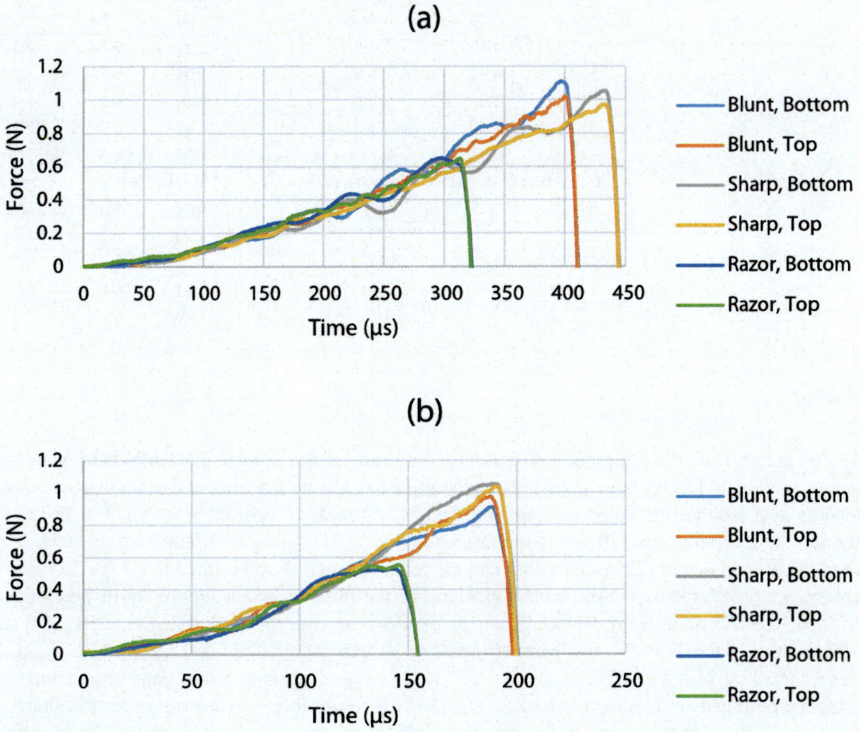

Fig. 14.7 Representative load cell traces for (**a**) 10 m/s impact velocity (475–507 s^{-1} approximate strain rate) and (**b**) 20 m/s impact velocity (975–1019 s^{-1} approximate strain rate)

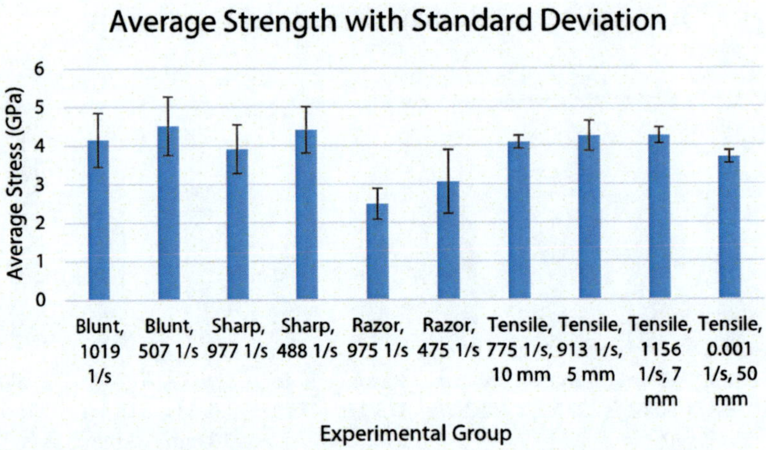

Fig. 14.8 Average stress comparison. Tensile values are from Ref. [3]

Table 14.1 Experimental data summary. Strain rate is included for correlation with velocity

Transverse impact			**Velocity (m/s)**		**Strain rate (s^{-1})**	**Strain (%)**		**Strength (GPa)**	
Geometry	**Speed**	**Sample size**	**Avg**	**SD**	**Avg**	**Avg**	**SD**	**Avg**	**SD**
Blunt	Fast	10	21.14	0.79	1018.6	1.94	0.34	4.15	0.70
	Slow	14	10.52	0.40	507.2	2.24	0.34	4.50	0.75
Sharp	Fast	21	20.27	0.99	976.8	2.04	0.36	3.91	0.63
	Slow	15	10.12	0.50	487.9	2.36	0.27	4.40	0.60
Razor	Fast	10	20.24	0.69	975.4	1.42	0.27	2.48	0.40
	Slow	16	9.86	0.41	475.1	1.55	0.57	3.06	0.83
Tensile data			**Gage length (mm)**		**Strain rate (s^{-1})**	**Strain (%)**		**Strength (GPa)**	
			Avg		**Avg**	**Avg**	**SD**	**Avg**	**SD**
			10		775	3.00	0.24	4.08	0.17
			5		913	2.58	0.31	4.54	0.39
			7		1156	3.51	0.57	4.25	0.21
			50		0.001	3.96	0.36	3.69	0.17

14.4 Conclusions

This study examines the effects of loading geometry and loading rate on the failure of UHMWPE Dyneema® SK76 single fibers by developing a modified Kolsky bar experimental technique to transversely impact the single fibers. For all geometries, an increased loading rate resulted in a decrease in both failure strain and apparent strength. For tests at approximately 1000 s^{-1} strain rates, the sharp indenter (20 μm radius) resulted in a 5.07% increase in failure strain and a 5.28% decrease in strength relative to the blunt indenter (200 μm radius) and the razor (2 μm radius) resulted in a 27% decrease in failure strain and a 40% decrease in strength relative to the blunt indenter. For the blunt indenter, failure strain was reduced by 43.5% at 507 s^{-1} and 51.0% at 1019 s^{-1} relative to quasi-static tensile experiments. For the blunt indenter at 507 s^{-1}, failure strain was reduced by 25.5% relative to the 775 s^{-1} tensile experiments at 10 mm gage length, and for the blunt indenter at 1019 s^{-1}, failure strain was reduced by 24.8% relative to the 913 s^{-1} tensile experiments at 5 mm gage length and 44.8% relative to the 1156 s^{-1} tensile experiments at 7 mm gage length. The slight increase in the failure strain for the sharp indenter at both nominal strain rates indicates that the radius is not small enough to induce total shear failure of the fiber like the razor, yet the geometry is influencing the failure mechanism such that the average strain can be increased, possibly at the expense of increased progressive damage through localized failure. For tests at approximately 500 s^{-1}, the sharp indenter demonstrated a 5.36% increase in failure strain and a 2.24% decrease in strength relative to the blunt indenter and the razor demonstrated a 24.53% decrease in failure strain and a 32.09% decrease in strength relative to the blunt indenter. For the blunt indenter, strength was increased by 22.1% at 507 s^{-1} and 12.5% at 1019 s^{-1} relative to quasi-static tensile experiments. For the blunt indenter at 507 s^{-1}, strength was reduced by 10.4% relative to the 775 s^{-1} tensile experiments at 10 mm gage length, and for

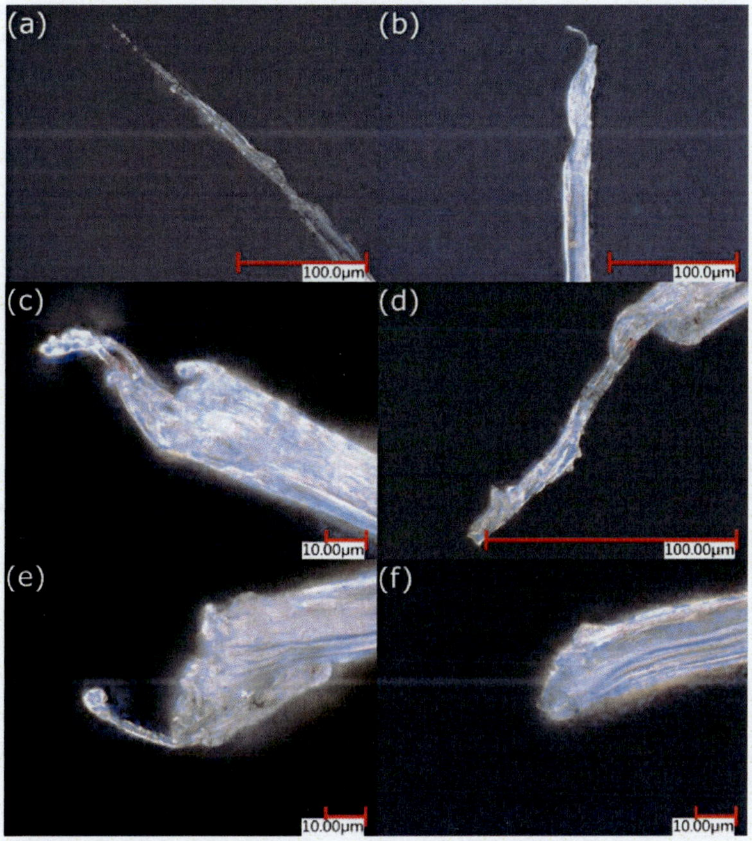

Fig. 14.9 Broken fiber ends for each experimental case: **(a)** blunt, 1018.6 s^{-1}; **(b)** blunt, 507.2 s^{-1}; **(c)** sharp, 976.8 s^{-1}; **(d)** sharp, 487.9 s^{-1}; **(e)** razor, 975.4 s^{-1}; **(f)** razor, 475.1 s^{-1}

the blunt indenter at 1019 s^{-1}, failure strain was reduced by 2.1% relative to the 913 s^{-1} tensile experiments at 5 mm gage length and 2.4% relative to the 1156 s^{-1} tensile experiments at 7 mm gage length. The consistency of the geometric effects when controlling for strain rate indicate the reliability of the results despite relatively large standard deviations. Therefore, the slight decrease in apparent strength in the sharp indenter experiments relative to the blunt indenter for both strain rates despite the increase in apparent strain further supports the conclusion that the indenter geometry is influencing the failure mode even though the differences in the strength and strain data are subtle.

The microscope imaging is useful in elucidating failure modes of the fibers. Blunt tests clearly undergo fibrillation similar to tensile tests. For razor loading, fibrils are much more likely to shear, which appears to occur at significantly smaller levels of deformation. Sharp tests, in contrast to both, demonstrate a combination of fibrillation and fibril shear. This phenomenon could explain the apparent mismatch between failure strain and strength in the case of the sharp experiments compared to the blunt experiments, where failure strain was increased slightly, yet strength was reduced. Initially impacted fibrils may undergo shear failure while fibrils that are not on the impact side fail later due to fibrillation. Strain localization may not be significant enough to result in early macroscopic fiber failure but the reduced load-carrying capacity would result in lower apparent failure stress. Further modeling efforts may yield more information regarding the stress-strain states, strain concentration, and the level of multiaxial loading experienced by the fibers. Future investigation using finite element modeling of the experiments is necessary to develop a strain rate–dependent multiaxial failure criterion [7].

Acknowledgments Research funding was provided through Oak Ridge Associated Universities. Equipment and facilities were provided by the Army Research Laboratory at Aberdeen Proving Ground, Maryland. The views and conclusions contained in this document are those of the authors and should not be interpreted as representing the official policies, either expressed or implied, of the Army Research Laboratory or the US Government. The US Government is authorized to reproduce and distribute reprints for Government purposes notwithstanding any copyright notation herein.

References

1. Krishnan, K., Sockalingam, S., Bansal, S., Rajan, S.D.: Numerical simulation of ceramic composite armor subjected to ballistic impact. Compos. Part B Eng. **41**(8), 583–593 (2010)
2. Sockalingam, S., Chowdhury, S.C., Gillespie, J.W., Keefe, M.: Recent advances in modeling and experiments of Kevlar ballistic fibrils, fibers, yarns and flexible woven textile fabrics – a review. Text. Res. J. **87**(8), 984–1010 (2017)
3. Sanborn, B., DiLeonardi, A.M., Weerasooriya, T.: Tensile properties of Dyneema SK76 single fibers at multiple loading rates using a direct gripping method. J. Dyn. Behav. Mater. **1**(1), 4–14 (2015)
4. Wang, H., et al.: Tensile properties of ultra-high-molecular-weight polyethylene single yarns at different strain rates. J. Compos. Mater. (2019)
5. Hudspeth, M., Li, D., Spatola, J., Chen, W., Zheng, J.: The effects of off-axis transverse deflection loading on the failure strain of various high-performance fibers. Text. Res. J. **86**(9), 897–910 (2015)
6. Sockalingam, S., Thomas, F.D., Casem, D., Gillespie, J.W., Weerasooriya, T.: Failure of Dyneema® SK76 single fiber under multiaxial transverse loading. Text. Res. J. (2018)
7. Thomas, F.D., Casem, D., Weerasooriya, T., Sockalingam, S., Gillespie Jr., J.W.: Influence of high strain rate transverse compression on the tensile strength of polyethylene ballistic single fibers. **1**, 339–344 (2019)
8. Smith, J.C., Fenstermaker, C.A., Shouse, P.J.: Stress-strain relationships in yarns subjected to rapid impact loading. Text. Res. J., 743–757 (1965)
9. Roylance, D.: Ballistics of transversely impacted fibers. Text. Res. J. **47**(10), 679–684 (1977)

Chapter 15
Using Crack Geometry to Determine Fracture Properties

Kimberley A. Mac Donald and Guruswami Ravichandran

Abstract Linear elastic fracture mechanics theory predicts a parabolic crack opening profile. However, direct observation of crack tip shape in situ for brittle materials is challenging due to the small size of the active crack tip region. By leveraging advances in optical microscopy techniques and using a soft brittle hydrogel material, we can measure crack geometry on the micron scale. For glasses and ceramics, expected crack opening displacements are on the order of nanometers. However, for hydrogels, we can achieve crack opening displacements on the order of hundreds of microns or larger while maintaining brittle fracture behavior. Knowing the elastic properties, we can use crack geometry to calculate the stress intensity factor, K, and energy release rate, G, during propagation. Assuming the gel is hyperelastic, we can also approximate the size of the nonlinear region ahead of the crack tip. Geometric measurement of fracture properties eliminates the need to measure complex boundary and loading conditions, allowing us to explore new methods of inducing crack propagation. Further, this allows us to define measures of fracture resistance in materials that do not fit the traditionally defined theories of fracture mechanics.

Keywords Brittle fracture · Confocal microscopy · Crack geometry · Polymers · Soft gels

15.1 Introduction

With advances in fields such as wearable devices and implants, soft robotics, and additive manufacturing, the structural performance of soft materials has become of great interest [1–3]. Materials that are mechanically compatible with biological tissues can help improve comfort of wearable devices and reduce inflammation, scarring, and rejection of implants [4]. Soft actuation and reducing joint friction are critical topics in robotics development [5]. Additionally, customized scaffolds, with the porosity to provide critical nutrients, are being used to grow cells into simple organs, where it has been shown that the scaffold's stiffness has a strong impact on cell growth [6–8].

In all these fields, the mechanical properties of soft polymers are of great interest. However, conventional gels exhibit very low toughness and brittle fracture properties, making them too delicate for most of these high demand applications [9–11]. While brittle fracture is well understood within the linear elastic regime for crystalline materials such as metals and ceramics, the microstructure of polymers is significantly different [12, 13]. Polymers consist of chains of monomers, typically hundreds to hundreds of thousands of monomer units in a chain, where strong covalent bonds form the backbone and weaker hydrogen bonds between chains combine with physical entanglement to form the polymer [14]. Crosslinking, such as in many polymer hydrogels, can provide further strong covalent bonds to form a networked polymer system. This means that mechanisms of polymer failure are significantly different from metals and ceramics. Where there are identifiable slip and cleavage planes in the crystalline structure of metals and ceramics, even networked polymers show complex chain interactions such as chain stretching and pull-out before the strong bonds between monomer units are affected and fracture occurs [15, 16].

Many polymers exhibit features of brittle fracture: no observable permanent deformation and smooth fracture surfaces perpendicular to the loading direction [17]. However, traditional measures of fracture properties were developed based on

K. A. Mac Donald (✉)
Mechanics of Materials, Sandia National Laboratories, Livermore, CA, USA
e-mail: kamacdo@sandia.gov

G. Ravichandran
Division of Engineering and Applied Science, California Institute of Technology, Pasadena, CA, USA
e-mail: ravi@caltech.edu

S. Xia et al. (eds.), *Fracture, Fatigue, Failure and Damage Evolution, Volume 3*, Conference Proceedings of the Society for Experimental Mechanics Series, https://doi.org/10.1007/978-3-030-60959-7_15

crystalline metals and ceramics. When applying these measures to polymers or other amorphous materials like glasses, the theories behind these measures no longer have the same meaning. However, the concept of fracture toughness as a measure of the energy needed to create new surfaces within the material is still applicable.

15.2 Background

Linear elastic fracture mechanics (LEFM) theory predicts a parabolic crack opening profile described as

$$\delta(r) = \frac{8K}{\sqrt{2\pi}E\lambda}\sqrt{r} \tag{15.1}$$

$$\text{where } \lambda = \begin{cases} 1 & \text{for plane stress,} \\ \frac{1}{1-\nu^2} & \text{for plane strain,} \end{cases} \tag{15.2}$$

E is Young's modulus, ν is Poisson's ratio, r is the horizontal distance along the crack surface away from the tip, and K is the stress intensity factor [17]. If we define δ_t as the crack opening where perpendicular lines from the vertex of the parabola (crack tip) intersect with the parabolic profile (crack edges), we can define $r_t = \delta_t/2$ as indicated in Fig. 15.1 [18, 19]. This allows us to determine K in terms of elastic constants and a measure of the crack tip opening displacement (CTOD, δ_t):

$$K = \frac{\sqrt{\pi\delta_t}E\lambda}{4}. \tag{15.3}$$

For small scale yielding, the energy release rate is then

$$G = \frac{K^2}{E\lambda} = \frac{\pi\delta_t E\lambda}{16}. \tag{15.4}$$

Many soft materials, including the gel used in this study, could be considered hyperelastic, which means there is a nonlinear region near the crack tip in which the parabolic crack profile would be altered. Geubelle and Knauss [20] estimate the size of this nonlinear region to be

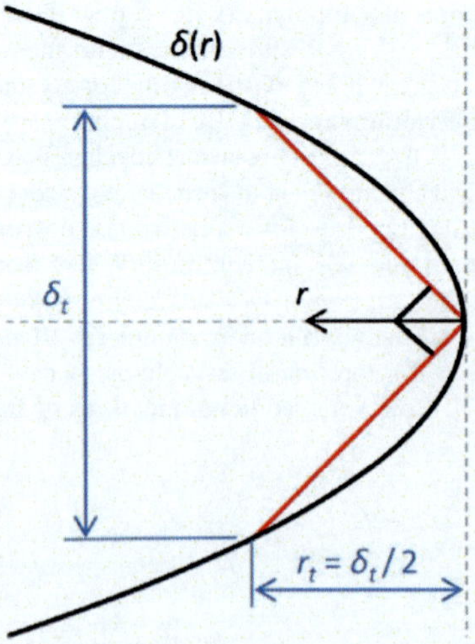

Fig. 15.1 Parabolic crack opening profile with crack tip opening displacement, δ_t, determined by the intersection of perpendicular lines from the crack tip with the crack edges

$$r_{NL} = \frac{1}{3\pi}\frac{J}{\mu} \tag{15.5}$$

where μ is the shear modulus, which can be related to E and ν for a linear elastic material, and J is Rice's J-integral, which corresponds to the energy release rate as $J = G$ for a linear elastic material [17, 19]. Thus, using Eq 15.4, r_{NL} is found to depend only on the Poisson's ratio (compressibility) and δ_t:

$$r_{NL} = \frac{1+\nu}{24}\lambda\delta_t. \tag{15.6}$$

For an incompressible hyperelastic gel, $\nu = 0.5$, so for plane stress $r_{NL} = \delta_t/16$, and for plane strain $r_{NL} = \delta_t/12$. In both cases, the nonlinear region is significantly smaller than the crack tip opening displacement, so assuming LEFM holds in determining the CTOD is appropriate.

15.3 Summary of Experiment and Results

Due to the high water content of many hydrogels (often >90%) and their very low toughness, it can be challenging to conduct traditional fracture tests on these materials. We also conduct in situ confocal microscope imaging on the submillimeter scale to understand both crack geometry and the damage processes occurring in the polymer network, introducing further experimental challenges. For these reasons, we use a fracking-inspired method to induce stable, slow crack propagation in a thin, confined disk of hydrogel. The experimental methods are described in detail in [21].

A second-order polynomial (parabola) fits the crack shapes observed in the experiments very well, as shown in Fig. 15.2. Details of the calculated fracture properties of this gel using this method to induce crack propagation can be found in [21]. For a propagating crack in this gel, which has a Young's Modulus of approximately 78 kPa, the mean CTOD (δ_t) is 339 μm

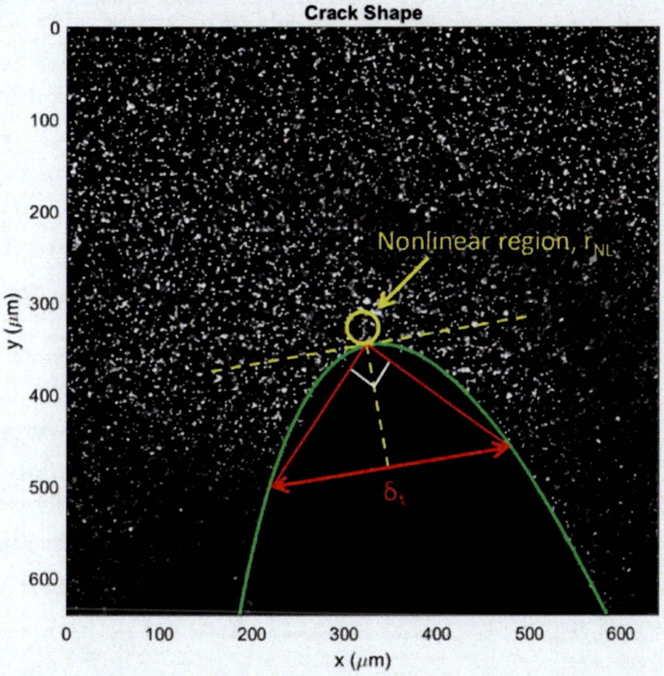

Fig. 15.2 A parabola fit to a confocal microscope image of an internal plane of the gel indicating the measured CTOD (mean value, 339 μm) and estimated size of the nonlinear region (mean value, 28 μm)

corresponding to a mean SIF (K) and ERR (G) of 850 $Pa\sqrt{m}$ and 6.9 J/m^2 respectively. The estimated size of the nonlinear region near the crack tip (r_{NL}) is 28 μm, which corresponds to about 22 pixels in the captured images, roughly the size of an appropriate digital image correlation subset for this speckle pattern. Errors in these calculated values are comparable to the errors reported for more traditional fracture experiments in similar soft gel materials [9].

Some features of crack renucleation in this networked polymer hydrogel were observed at the submillimeter scale such as intact ligaments and indicators of local toughening. However, to better understand nonlinear processes in this gel, higher resolution imaging of the crack tip nonlinear region is needed.

In this study, we were able to use crack geometry and the polymer's material properties to determine traditional measures of fracture toughness without needing to define and measure complex boundary conditions. We were also able to determine the scale at which we would expect nonlinear processes to affect the crack.

Acknowledgments This material is based upon work supported by the National Science Foundation Graduate Research Fellowship under Grant No. DGE-1144469 and the Designing Materials to Revolutionize and Engineer our Future (DMREF) program under Award No. DMS-1535083. Imaging was performed in the Biological Imaging Facility, with the support of the Caltech Beckman Institute and the Arnold and Mabel Beckman Foundation.

Sandia National Laboratories is a multimission laboratory managed and operated by National Technology & Engineering Solutions of Sandia, LLC, a wholly owned subsidiary of Honeywell International Inc., for the US Department of Energy's National Nuclear Security Administration under contract DE-NA0003525.

References

1. Baldi, F., Bignotti, F., Peroni, I., Agnelli, S., Ricco, T.: On the measurement of the fracture resistance of polyacrylamide hydrogels by wire cutting tests. Polym. Test. **31**(3), 455–465 (2012)
2. Gong, J.P., Osada, Y.: Soft and wet materials: from hydrogels to biotissues. In: Cloitre, M. (ed.) High Solid Dispersions, pp. 203–246. Springer, Berlin/Heidelberg (2010)
3. Long, R., Hui, C.Y.: Fracture toughness of hydrogels: measurement and interpretation. Soft Matter. **12**(39), 8069–8086 (2016)
4. Moshayedi, P., Ng, G., Kwok, J.C.F., Yeo, G.S.H., Bryant, C.E., Fawcett, J.W., Franze, K., Guck, J.: The relationship between glial cell mechanosensitivity and foreign body reactions in the central nervous system. Biomaterials. **35**(13), 3919–3925 (2014)
5. Coyle, S., Majidi, C., LeDuc, P., Hsia, K.J.: Bioinspired soft robotics: material selection, actuation, and design. Extrem. Mech. Lett. **22**, 51–59 (2018)
6. Engler, A.J., Sen, S., Sweeney, H.L., Discher, D.E.: Matrix elasticity directs stem cell lineage specification. Cell. **126**(4), 677–689 (2006)
7. Whang, M., Kim, J.: Synthetic hydrogels with stiffness gradients for durotaxis study and tissue engineering scaffolds. Tissue Eng. Regen. Med. **13**(2), 126–139 (2016)
8. Yeung, T., Georges, P.C., Flanagan, L.A., Marg, B., Ortiz, M., Funaki, M., Zahir, N., Ming, W., Weaver, V., Janmey, P.A.: Effects of substrate stiffness on cell morphology, cytoskeletal structure, and adhesion. Cell Motil. **60**(1), 24–34 (2005)
9. Agnelli, S., Baldi, F., Bignotti, F., Salvadori, A., Peroni, I.: Fracture characterization of hyperelastic polyacrylamide hydrogels. Eng. Fract. Mech. **203**, 54–65 (2018)
10. Nilsson, F.: A tentative method for determination of cohesive zone properties in soft materials. Int. J. Fract. **136**(1), 133–142 (2005)
11. Tanaka, Y., Shimazaki, R., Yano, S., Yoshida, G., Yamaguchi, T.: Solvent effects on the fracture of chemically crosslinked gels. Soft Matter. **12**(39), 8135–8142 (2016)
12. Ahmed, E.M.: Hydrogel: preparation, characterization, and applications: a review. J. Adv. Res. **6**(2), 105–121 (2013)
13. Denisin, A.K., Pruitt, B.L.: Tuning the range of polyacrylamide gel stiffness for mechanobiology applications. ACS Appl. Mater. Interfaces. **8**(34), 21893–21902 (2016)
14. Odian, G.: Principles of Polymerization. Wiley InterScience Electronic Collection. Wiley, Hoboken (2004)
15. Brighenti, R., Spagnoli, A., Carpinteri, A., Artoni, F.: Notch effect in highly deformable material sheets. Thin-Walled Struct. **105**, 90–100 (2016)
16. Naficy, S., Brown, H.R., Razal, J.M., Spinks, G.M., Whitten, P.G.: Progress toward robust polymer hydrogels. Aust. J. Chem. **64**(8), 1007–1025 (2011)
17. Anderson, T.L.: Fracture Mechanics: Fundamentals and Applications, 3rd edn. CRC Press, Boca Raton (2005)
18. Rice, J.R.: A path independent integral and the approximate analysis of strain concentration by notches and cracks. J. Appl. Mech. **35**(2), 379–386 (1968)
19. Shih, C.F.: Relationships between the J-integral and the crack opening displacement for stationary and extending cracks. J. Mech. Phys. Solids. **29**(4), 305–326 (1981)
20. Geubelle, P.H., Knauss, W.G.: Finite strains at the tip of a crack in a sheet of hyperelastic material: I. Homogeneous case. J. Elast. **35**, 61–98 (1994)
21. Mac Donald, K.A., Ravichandran, G.: Crack propagation and renucleation in soft brittle hydrogels. Int. J. Fract. **222**, 37–52 (2020)

Chapter 16
Dynamic Fracture-Toughness Testing of a Hierarchically Nano-Structured Solid

Kyung-Suk Kim, Hanxun Jin, Tong Jiao, and Rodney J. Clifton

Abstract The polyurea coating is found very useful in strengthening structures ranging from helmets to concrete structures under impact or blast loading. We believe that the hierarchical architecture of nano and microstructures is the bases of the strengthening mechanism, which provides scale-dependent stress laxation and energy dissipation. Here, a challenge is to characterize the strengthening mechanisms not only in the bulk of the copolymer but also at the coating/substrate interface. To this end, we have found that the tapping-mode images of an atomic-force-microscope (AFM) are ideal markers for digital image correlation (DIC) analysis of nano/micro-scale deformation. The tapping-mode images typically exhibit clustered hierarchical structures of hard and soft domains that can trace multiscale deformation mechanisms. To study the role of the hierarchical deformation mechanisms in dynamic toughening, we have developed a line-image shearing interferometer (L-ISI) for plate impact experiments of dynamic fracture testing. The L-ISI measures the variation of the normal-displacement-gradient over time along a line on the back surface of a pre-cracked specimen loaded by plate impact. The time history of the displacement gradient forms fringes on the streak-camera image, and the fringes are inverted to determine the time history of the crack speed and the dynamic toughness.

Keywords Polyurea · Dynamic toughness · Line-image shearing interferometer · Tapping-mode AFM DIC

16.1 Introduction

Recently, a nanophase-segregated elastomeric copolymer, polyurea, has gained significant attention on its nano- and micrometer scale mechanisms of dynamic fracture toughening [1, 2]. Bulk polyurea is typically self-assembled to be clustered in hard isocyanate group and soft diamine group phases at nanometer-length scales, and not homogeneous but segregated. As a consequence, polyurea is known to contain a two-phase microstructure consisting of discrete, rod-shaped, the so-called hard domains which are dispersed within a continuous soft matrix. Figure 16.1a shows bright AFM images of 5–10 nm diameter rods of hard isocyanate group embedded in the soft diamine group matrix. The high-resolution image was obtained with a very sharp AFM tip of 2 nm radius, in a tapping mode. However, when a relatively blunt AFM tip of ~20 nm radius is used with noncontact mode, hierarchical clustering of the hard and soft phases with ~200 nm periodicity is revealed, as shown in Fig. 16.1b.

The segregated and hierarchically clustered nanostructures are believed to provide peculiar dynamic deformation mechanisms responsible for ultra-high strength of polyurea under high strain-rate and/or pressure loading. For studying the dynamic response of polyurea, much research on the high-strain-rate response has addressed its deformation under dynamic loading without tackling its resistance to failure associated with failure-initiation and crack-propagation processes. However, the survivability of the structural element that the coating is intended to protect depends on its adhesion to the structural component, and the development of a mechanism-based understanding of failure in elastomers and their interfaces with other materials is critically important.

K.-S. Kim (✉) · H. Jin
Nano and Micromechanics Laboratory, School of Engineering, Brown University, Providence, RI, USA
e-mail: kyung-suk_kim@brown.edu; hanxun_jin@brown.edu

T. Jiao · R. J. Clifton
Plate Impact Facility, Prince Laboratory, School of Engineering, Brown University, Providence, RI, USA
e-mail: Tong_Jiao@brown.edu; rodney_clifton@brown.edu

S. Xia et al. (eds.), *Fracture, Fatigue, Failure and Damage Evolution, Volume 3*, Conference Proceedings of the Society for Experimental Mechanics Series, https://doi.org/10.1007/978-3-030-60959-7_16

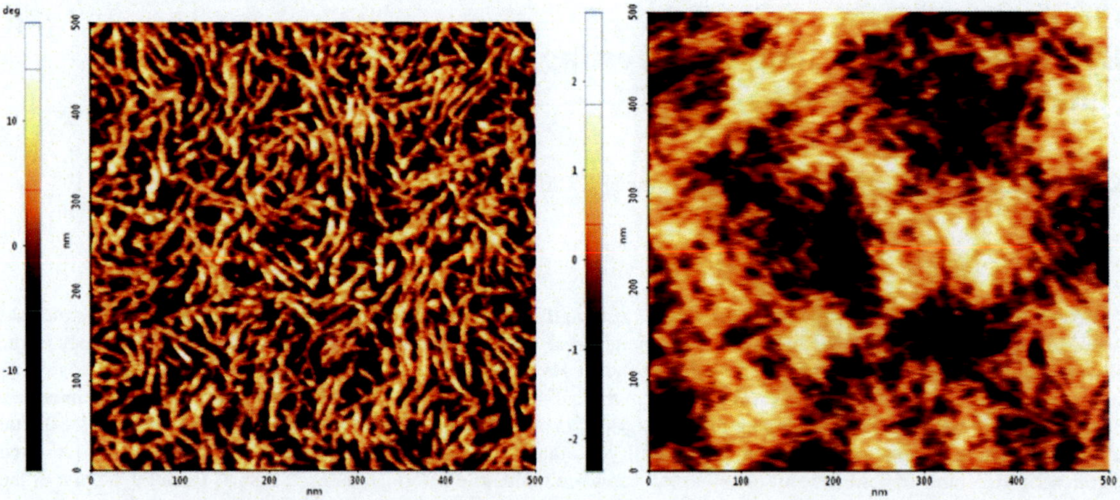

Fig. 16.1 (**a**) A sharp-tip AFM tapping-mode image and (**b**) a blunt-tip AFM noncontact-mode image of Polyurea PU1000

16.2 Background

Regarding the failure analysis, initiation of tensile failure in polyurea PU1000 under conditions of uniaxial strain has been observed through plate-impact, release-wave experiments [3], which are often used for spall tests. In these experiments, a thin flyer impacts a test-sample plate to create a compressive pulse that reflects from the rear surface of the polyurea and generates a tensile pulse after the reflected pulse passes through the oncoming compressive pulse. The dynamic plane-wave test revealed tensile failure stress of polyurea PU1000 approximately 105 MPa under conditions of uniaxial strain. However, polyurea did not spall at this stress level within the tensile-pulse duration. Instead, microvoids opened on the putative "spall plane," which continued to transmit tensile stress. Knowing the failure initiations stress, actual constitutive relations of the full failure process in spalling require realistic fracture models involving void initiation, growth, and coalescence. Such failure processes are often characterized by cohesive failure models that provide a computational foundation for modeling failure in elastomers.

The development of such cohesive models appears to require results of more well-characterized crack-propagation experiments than "spall" tests that involve heterogeneous nucleation and growth of voids. Such experiments, involving plane wave loading of plates with a single mid-plane crack have been previously developed and used to study crack growth in a metal, AISI 4340, under mode I as well as II & III loading conditions [4]. These plane wave experiments are attractive because of their simplicity in tracing the crack-tip loading conditions through measurements of the motion at the traction-free rear surface of the pre-cracked target plate. However, all previous studies used interferometric techniques that could only measure velocity-time profiles at a single point on the rear surface of the plate. The single-point measurement was particularly troublesome in its accuracy to inversely assess the crack-tip loading profiles.

16.3 Analysis

Now, a high-speed streak camera is available, and it is possible to monitor motion—along a line—instead of at a single point at the rear surface. Here, we present a line-image shearing interferometer (L-ISI) for plate impact experiments of dynamic fracture testing. The L-ISI optical circuit is shown schematically as red lines in Fig. 16.2a. In the optical circuit, a laser beam is modulated to synchronously switch the beam by an acousto-optic modulator (AOM) located at the focal plane of the spherical lens SL2, for the exposure duration of a high dynamic range streak camera (C7700). Then the laser light is collimated and aligned by two cylindrical lenses CL1 and CL2 to make a slit illumination on the mirror-finished rear surface of the specimen. The orientation of the illuminated slit is perpendicular to the crack-front line. Then, the slit image of the reflected light is projected on the image plane of the streak camera through the two spherical lenses SL2 and SL3. At the

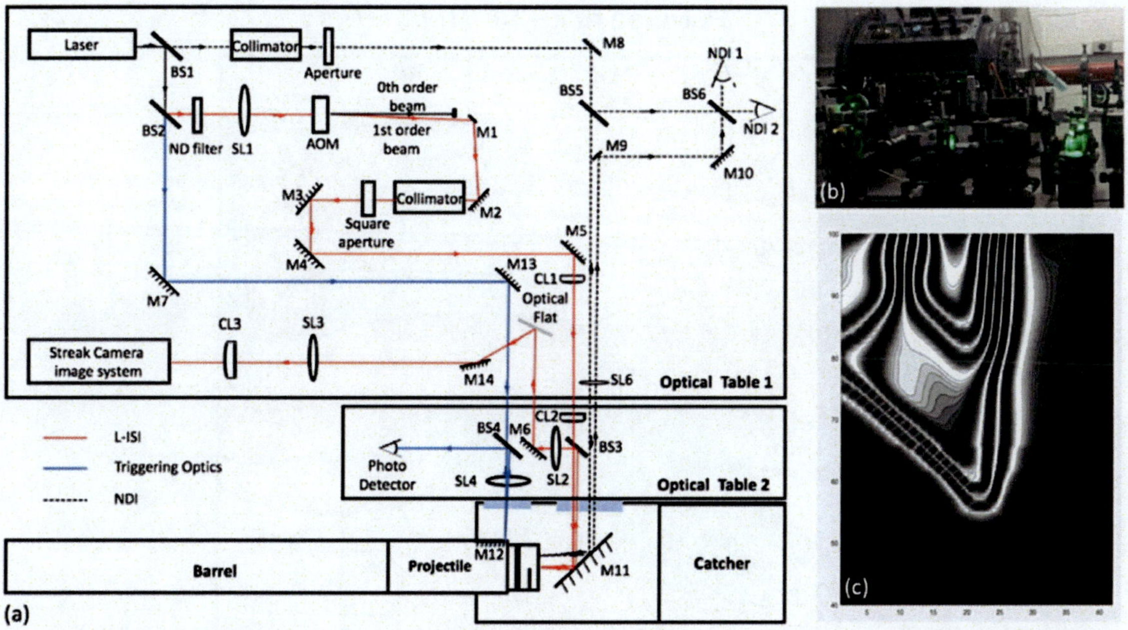

Fig. 16.2 (**a**) A line-image shearing interferometer (L-ISI) with NDI; (**b**) plate-impact chamber; (**c**) an FEM-simulation of L-ISI fringe

focal point of SL2, an optical flat of 150 μm thickness is used to provide an image shearing of 50 μm at the object plane. The slit length is 22 mm and the slit width ~ 2 mm at the object plane. The center-cracked polyurea specimen has dimensions of 50 mm × 50 mm × 4.45 mm. The optical circuit of the L-ISI can accommodate the fringe density and frequency for streak camera C7700 up to 400 m/sec velocity of symmetric impact, that is, the flyer is made of the same polyurea with the half-thickness of the specimen.

Figure 16.2a also shows an optical triggering circuit for the streak camera with blue lines and a normal displacement interferometer circuit with black dotted lines. Since the L-ISI provides the time variation of the displacement gradient along a line, a time trace of the displacement at a point has to be measured to get full information of the displacements in the position and time coordinate. Figure 16.2b exhibits a photograph of the laser optical setup aligned with the impact vacuum chamber. Figure 16.2c displays finite element simulation of the L-ISI fringes for an impact velocity of 200 m/sec. The vertical axis represents time in μs and the horizontal axis position in mm. The initial crack is located at the center of the position coordinate. The fringes develop upward once the tensile wave arrives at the crack plane. Then, the nearly symmetric fringes become significantly asymmetric when the crack begins to propagate to the left direction, after an incubation time. Subsequently, the fringes show crack arrest and restart as the stress wave reverberates back and forth in the specimen plate.

We have carried out a shot with an impact speed of 164 m/sec. We missed the time window for L-ISI; however, we could get the incubation time of crack-growth initiation, ~1 μs, and the growing distance, ~400 μm, of its first arrest. A finite element simulation with a triangular cohesive zone model of two parameters – the cohesive strength, 220 MPa, and the failure separation, ~50 μm– that match the incubation time and the growing distance gives an estimation of the toughness, 5500 J/m^2. On the other hand, the linear elastic fracture mechanics (LEFM) estimation of the toughness provides 6470 J/m^2, which overestimates ~18%. The LEFM estimation of the fracture toughness is obtained by $K_I(t) = n(\nu)\,\sigma_0 C_l^{1/2} t^{1/2}$, where K_I is the fracture toughness, σ_0 (~150 MPa) the stress of the tensile pulse, C_l (~2100 m/sec) the longitudinal wave speed, t (~1 μs) the incubation time, and $n(\nu) = 2\sqrt{(1-2\nu)/\pi}/(1-\nu)$ with Poisson's ratio ν (~0.41). The rate of crack-tip loading is estimated $\dot{K} \sim 10^7 \mathrm{MPa}\sqrt{\mathrm{m}}/\mathrm{sec}$. The toughness is given by $G_C = \left(1-\nu^2\right)K_{IC}^2/E$ (~6470 J/m2) with E~1.25GPa. The strength and the toughness are displayed on an Ashby map [5] in Fig. 16.3. Indeed, the toughness is also found high for the very high strength at the high rate of loading.

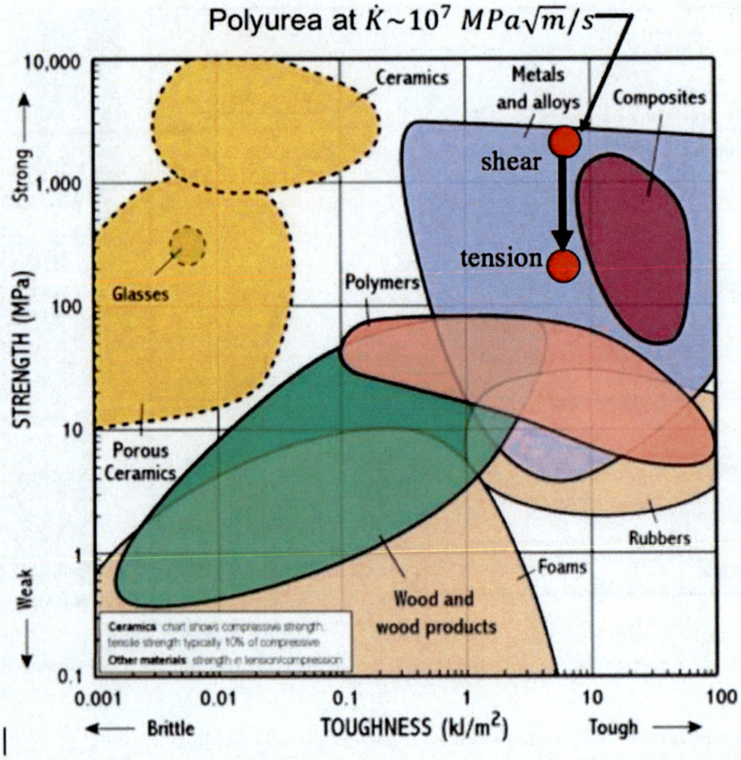

Fig. 16.3 High toughness of polyurea with reduce strength in tension at high rate of loading is plotted on an Ashby map [5]

16.4 Conclusion

Dynamic toughness of polyurea PU1000 is measured with a newly developed L-ISI, applying plane-wave loading of a center-cracked specimen with plate impact. The toughness is measured 5500 J/m^2 under a very high rate of crack-tip loading, $\dot{K} \sim 10^7 \text{MPa}\sqrt{\text{m}}/\text{s}$. The toughness is relatively high. The strength under high strain rate is found substantially lower, ~220 MPa, at a tensile loading than that, ~2GPa, at a compressive shear loading. However, polyurea PU1000 has enough cohesive ductility, ~50 μm, to exhibit high toughness under the reduced strength in tension. This result shows that the high nominal strength of polyurea at high strain-rate under high pressure, ~2GPa, is mainly due to high hydrostatic pressure. The underlying strengthening mechanisms under a wide range of hydrostatic pressure are to be further investigated with the newly developed tapping-mode AFM DIC [6].

Acknowledgments This work is supported by grant N00014-18-1-2513, from the US Office of Naval Research.

References

1. Grujicic, M., et al.: Experimental and computational study of the shearing resistance of polyurea at high pressures and high strain rates. J. Mater. Eng. Perform. **24**(2), 778–798 (2014)
2. Kim, K.-S.: Pressure- and rate-dependent plastic flow of nano-phase segregated polyurea copolymer (IMECE Paper # -13559). Drucker Medal Symposium, ASME 2019 IMECE (November 2019)
3. Jiao, T., et al.: Pressure-Sensitivity and Tensile Strength of an Elastomer at High Strain Rates. In: Elert, M., Furnish, M.D., Chau, R., Holmes, N., Nguyen, J. (eds.) Shock Compression of Condensed Matter, pp. 707–710 (2007)
4. Ravichandran, G., Clifton, R.J.: Dynamic fracture under plane wave loading. Int. J. Fracture. **40**, 157–201 (1989)
5. Bolton, W., Higgins, R.A.: Materials for Engineers and Technicians. Newnes, London (2010)
6. Kim, K.-S.: Digital Image Correlation at a Small World of Nano Science and Technology (Plenary). iDICs Annual Conference, October 2019